Noor Kareem Jumaa
Firas Ali Sabir

Quantum Key Distribution Using FPGA: A Real Time Prototype, Design, and Analysis

Noor Kareem Jumaa
Firas Ali Sabir

Quantum Key Distribution Using FPGA: A Real Time Prototype, Design, and Analysis

Noor Publishing

Imprint

Any brand names and product names mentioned in this book are subject to trademark, brand or patent protection and are trademarks or registered trademarks of their respective holders. The use of brand names, product names, common names, trade names, product descriptions etc. even without a particular marking in this work is in no way to be construed to mean that such names may be regarded as unrestricted in respect of trademark and brand protection legislation and could thus be used by anyone.

Cover image: www.ingimage.com

Publisher:
Noor Publishing
is a trademark of
Dodo Books Indian Ocean Ltd., member of the OmniScriptum S.R.L Publishing group
str. A.Russo 15, of. 61, Chisinau-2068, Republic of Moldova Europe
Printed at: see last page
ISBN: 978-620-4-72006-7

Quantum Key Distribution Using FPGA: A Real Time Prototype, Design, and Analysis

By

Noor Kareem Jumaa
Dr. Firas Ali Sabir

a

About This Book

This book is an updated copy of a thesis submitted to the college of engineering in the University of Baghdad in partial fulfillment of the requirements for the degree of Master of Science in electronics and communications / computer engineering. It presents a study and analysis to investigate of a secret-key sharing in quantum technique and try to increase the key length and security level in the sifting key stage. Enhancements in sifting key stage and error correction stage are done by repeating the QKD protocol and encoding the key bits before sending them by adding a C_QUBITS technique. After that, 60% base probability is used to increase the protocol gain.

Also, Field Programmable Gate Array (FPGA) is used in the proposed designed system to implement the generators to both the base generator and the random key bits by applying the Linear Feedback Shift Register (LFSR). LFSR algorithm is implemented in FPGA using Very High Speed Integrated Circuit Hardware Description Language (VHDL). This design is implemented with the default FPGA clock period which is 20 ns; the performance analysis shows a good enhancement in the speed of generation which is less than 20 ns and the number of used hardware pins is just 3 pins. The measurements in this book are done using Matlab.

Table of Abbreviations

ABER	Algorithm Bit Error Rate
AMS	Analogue and Mixed-Signals
ASIC	Application Specific Integrated Circuit
BB84	Bennett-Brassard 1984
BER	Bit Error Rate
bps	Bit per Second
CLB	Configurable Logic Block
EEPROM	Electrical Erasable Programmable Read-Only Memory
EPROM	Erasable Programmable Read Only Memory
FPGA	Field Programmable Gate Array
HDL	Hardware Description Language
IC	Integrated Circuit
IOB	Input/output Blocks
LFSR	Linear Feedback Shift Register
LUT	Lookup Table
ML	Multi Level
PC	Personal Computer
QKD	Quantum Key Distribution
Qubit	Quantum Bit
SPb	Second per bit
SRAM	*Static Random-Access Memory*
UDP	User Datagram Protocol
VHDL	(Very high speed integrated circuit) Hardware Description Language

Table of Symbols

$\lvert \Psi \rangle$	Qubit state
$\lvert 0 \rangle$	Zero state
$\lvert 1 \rangle$	One state
α	Simple amplitude of $\lvert 0 \rangle$ or $\lvert 00 \rangle$
β	Simple amplitude of $\lvert 1 \rangle$ or $\lvert 01 \rangle$
∂	Simple amplitude of $\lvert 10 \rangle$
λ	Simple amplitude of $\lvert 11 \rangle$
$\lvert \updownarrow \rangle$	Polarization with 90º
$\lvert \leftrightarrow \rangle$	Polarization with 0º
$\lvert \backslash \rangle$	Polarization with -45º
$\lvert / \rangle$	Polarization with 45º
(+)	Rectilinear base
(×)	Diagonal base
H	Hadmard space
U	Unitary matrix of Hadmard space
$\otimes$	Tensor product
H_n	n-dimensional Hilbert space (H_n)
I	Pauli quantum gate
X	Pauli _X (NOT) quantum gate
Y	Pauli_Y quantum gate
Z	Pauli_Z quantum gate
S	Phase _S quantum gate
T_F	Toffoli gate
F_R	Fredkin gate
g_p	Protocol gain
δ	Base probability
P	Probability error rate
e	Number of errors
r	Length of key subset
q	Base probability of key subset
P_1	Probability error rate of stage 1
P_2	Probability error rate of stage 2

P	Average probability error rate
e_1	Estimated number of errors of first raw key subset.
e_2	Estimated number of errors of second raw key subset.
r_1	Length of the first subset of raw key.
r_2	Length of the second subset of raw key.

Contents

List of Tables

Chapter One

Introduction

1.1　General Overview

In spite of the fact that many research is still in its infancy, quantum phenomena have many promising ideas for applications that have started to develop. One important application of this technology is **quantum cryptography**, where the exchange of individual quantum systems between two correspondents enables them to establish a shared random key. The field of quantum cryptography was pioneered by Wiesner Around 1970; his research was not published until 1983. In 1979, Bennett and Brassard, who knew about Wiesner's ideas, realized that they could be used as a substitute for public-key cryptography. Two users, who shared no secret initially can communicate secretly with absolute and provable security. Their early quantum cryptographic schemes, BB84 protocol developed between 1982 and 1984. Experimental quantum key distribution was demonstrated for the first time in 1989. It was published only in 1992 by Bennett and Brassard. Since then, tremendous progress has been made. Today, QKD protocols have been implemented outside the laboratory. For example, china State Key Laboratory of Integrated Services Networks, implemented a practical QKD system over a 120 Km optical fiber optical network [1, 2, 3].

1.2 Motivation

Quantum computers rely on quantum computation which is the technology of the future [4]. The basic unit which the quantum computing processes is the quantum bit (qubit) which can exist in states taking a binary value of 0 or 1 and it can be in a superposition state which takes a value between 0 and 1 according to the probability of the 0 state and the probability of 1 state. Thus, quantum computing relies on probability. It should be noticed that quantum computers are prone to make errors and this opens a large area of search in quantum field to solve many problems [5, 6].

Traditional key distribution methods have security, efficiency, and gain problems. This thesis deals with solving these problems by making enhancements in key distribution protocol using quantum technique.

Nowadays, quantum cryptography is the most advanced and powerful cryptographic technique, that provides a complete security for two authorized parities to communicate over a private channel to exchange an enciphering random key. It is only used for key distributing and not for transmitting any useful data [7, 8].

Quantum cryptography is a method for Quantum Key Distribution (QKD), which provides a protection against the eavesdroppers by depending on the law of physics. The quantum cryptography depends on several important basics of quantum mechanics, no cloning theorem, qubits, the principle of photon polarization and the Heisenberg Uncertainty principle [9].

No cloning theorem prevents a third unauthorized party from decoding the transmitted information. Thus, if an eavesdropper tries to catch the information, the eavesdropper can not do this without making a disturbance on the system. So, both sender and receiver parties detect the eavesdropping action. [10]

The photon polarization principle states that unknown quantum state (qubit) can not be copied by an eavesdropper due to no-cloning theorem [10].

The Heisenberg uncertainty depends on quantum theory that states, a certain pair of physical properties are complementary. Thus, if one property is measured, the other is changed; so, any attempts to monitor the quantum state of a photon will change the properties and will be detectable and disturb the quantum system [10].

1.3   Literature Survey

To show the prime importance of Quantum Cryptography, some of relevant works in the field need to be reviewed.

- ❖ In 1999, Tayeh achieved a simulated quantum key distribution system and approved the security of this system over several type of network communication. Errors were discovered and discarded. An important mathematical technique called privacy amplification was implemented and eavesdropper information about the final cryptographic key was reduced to an arbitrarily value smaller than one bit [1].

- ❖ In 2002, Portland Quantum Logic Group developed complete methodologies, software tools and circuits for quantum logic and emulated the quantum circuits using standard reconfigurable FPGA technology and FPGA-based evolvable quantum hardware and they accelerated the quantum gate processes over Xilinx FPGA kit [3].

- ❖ In 2002, Navez and Assche proposed a quantum cryptography system by applying quantum computation using classical computer and improved the confidentiality of quantum cryptography over fiber optic communication [4].

❖ In 2007, Al-Daoud compared two well known quantum key distribution protocols: BB84 and SARG04 with many criteria such as photon number splitting attack, number of discarded bits, practical random key distribution, and practical maximum achievable distance. Using attenuated laser Source and single photon source he found that SARJ04 was better than BB84 in security issues [8].

❖ In 2008, Ware proposed a methodology for the development of a new modeling tool for quantum protocols and improved their weakness by improving the bit correctness to reach 0.854 probability [9].

❖ In 2009, Muhammad and Zukarnain discussed the implementation of QKD protocol (BB84 protocol), which is widely used today, simulated the quantum cryptography system with the existence of the eavesdropper and improved the final key length against the initial key length [11].

❖ In 2011, Kaur et al. made a comparison between classical cryptography and quantum cryptography and proved that the security level of quantum cryptography (quantum key distribution) is higher than classical cryptography [12].

❖ In 2011, Singh and Sharma proposed a mechanism which combined BB84 protocol at two levels; at the sender-receiver level and at the receiver-sender level; both levels were based on logic gates to reduce the probability of eavesdropping [13].

❖ In 2011, Devi and Raj implemented a quantum cryptography system based on BB84 protocol and proved that BB84 protocol has a Quantum Bit Error Rate (QBER) lower than other quantum cryptography protocols [14].

❖ In 2012, Kumar et al. implemented 8 bits, 16 bits, and 32 bits LFSR algorithm and studied and analyzed FPGA performance, they found that they had relatively high gate density, short design cycle, and low cost [15].

1.4   Aim of the Thesis

Many points are considered as a target for this work. The main points can be summarized as follows:

1. Performance enhancement of Quantum Cryptography.
2. The implementation of QKD over Altera Cyclone IV E FPGA.

1.5   Thesis Organization

As a necessary background, the basic concepts of quantum computations and quantum cryptography are the major concern of chapter one. In this section, each chapter contents, aims, and the work performed will be briefly shown.

❖ **Chapter Two** gives introduction to quantum computing and quantum principles with their important theories required in quantum circuits analysis. The chapter also deals with quantum gates and shows the structures of these gates. Also, Field Programmable Gate Array (FPGA) will be briefly introduced.

❖ **Chapter Three** gives the problems in classical cryptography. Quantum cryptography and the main quantum key distribution are discussed with its important protocols. This chapter also deals with the quantum attacking kinds and photon polarization.

❖ **Chapter Four** contains the proposed model of BB84 QKD protocol using Altera DE2-115 Cyclone IV E FPGA kit.

❖ **Chapter Five** contains the obtained results and their discussions.

❖ **Chapter Six** includes the conclusions and recommendations for the future works.

Chapter Two

Introduction to Quantum Computing and FPGA

2.1 Theoretical Background

Quantum computing field is a mixture of the computer science and quantum mechanics laws. This field evolved rapidly. Quantum computing is able to manipulate and communicate the quantum data. As the classical computing is able to deal with its classical data, quantum computing deals with its quantum data [16].

Field Programmable Gate Arrays (FPGAs) are pre-fabricated silicon devices that can be programmed electrically to become almost any kind of digital circuit or system. A number of compelling advantages are provided using FPGAs over fixed-function Application Specific Integrated Circuit (ASIC) technologies such as standard cells [17].

2.2 Quantum Computing and Quantum Computer

Quantum computing is to solve the problems that are intractable on digital computers to decrease the computational time for some problems by many orders of magnitude. The main advantage of quantum computation is the rapid parallel execution of logic operations achieved by using superposition states (entangled states); the entanglement seems to be natural for solving synchronization problems between distant agents [18].

It should be mentioned that quantum computing is not deterministic like classical computing; quantum computing is probabilistic and that quantum algorithms have to deal with the reality of measurement errors since in probabilistic computing, results of computation cannot be determined correctly every time a measurement of the result is made. But over all, quantum computing offers immense speedup in performing tasks such as data encryption and searching [19, 20].

A quantum computer is based on quantum computation. Quantum computers have the same building block as the classical computer, quantum bit which store the quantum information and quantum gate which processes the quantum bits. Quantum bit presents the smallest store unit and it is the basic unit which quantum computing processes it [21].

2.2.1 Quantum Computing Principles

Many parts of quantum mechanics are important for quantum computing. In this section, the main important parts are mentioned:

- Qubits
- Photon polarization
- Heisenberg uncertainty
- No cloning theorem
- Photon entanglement

(i)   Qubits

Quantum bit (qubit) is the quantum computer building unit which performs the basic variable used in quantum computing.

Recall that a classical bit can be in one of two states, either 0 or 1. A qubit can exist in states that take a binary value of $|0\rangle$ or $|1\rangle$ and it can be in a superposition state which takes a value between 0 and 1 according to the probability of the 0 state and the probability of 1 state according to Eq. (2.1) [22]:

$$|\Psi\rangle = \alpha|0\rangle + \beta|1\rangle \tag{2.1}$$

Where α and β are called probability amplitude or simple amplitude of $|0\rangle$ and $|1\rangle$, such that:

$$|\alpha|^2 + |\beta|^2 = 1 \tag{2.2}$$

The superposition state $|\Psi\rangle$ is a combination of states $|0\rangle$ and $|1\rangle$ and it performs the quantum computation power. That is, if there are two possible states, the system can be said to exist in both at once until its state is actually measured. Suppose there are two qubits; if these were two classical bits, then there would be four possible states: 00, 01, 10, and 11. Correspondingly, two qubits have four quantum states; they can be in any of these states or in a superposition of these state: $|\Psi\rangle = \alpha|00\rangle + \beta|01\rangle + \partial|10\rangle + \lambda|11\rangle$ where

$$|\alpha^2| + |\beta^2| + |\partial^2| + |\lambda^2| = 1 \tag{2.3}$$

Qubits are present with Ket notation as $|0\rangle$ and $|1\rangle$, which can be represented by vector form as shown:

$|0\rangle = \begin{bmatrix} 1 \\ 0 \end{bmatrix}$ and $|1\rangle = \begin{bmatrix} 0 \\ 1 \end{bmatrix}$ where $|0\rangle$ is being represented by an up-spin while $|1\rangle$ by a down-spin [3, 7, 16].

(ii)   Photon Polarization

Hilbert space is used to present the quantum bits as polarized photons according to some defined bases. The photon polarization principle describes how the photons of light can be polarized in an exact direction which prevents the eavesdropper from truly measuring the quantum states. Hilbert space can have and define various possible bases: the rectilinear basis ($| \updownarrow \rangle$, $| \leftrightarrow \rangle$), diagonal basis ($| \setminus \rangle$, $| / \rangle$) [5].

The quantum polarized bits, could be send at one of the basis states $|0\rangle$ or $|1\rangle$. The probability that a measurement of a qubit in $|0\rangle$ state is $|\alpha|^2$, and the probability that a measurement of a qubit in $|1\rangle$ state is $|\beta|^2$. The absolute value for the probabilities is needed since they are complex quantities.

When the quantum bits perform as described above, they will be sent as photon after photon; i.e., single photon transmission.

Photons also could be sent as pairs, where the pairs of qubits are capable to represent four distinct Boolean states, $|00\rangle$, $|01\rangle$, $|10\rangle$, $|11\rangle$, as well as all possible superposition of the states. This property is known as entanglement [4].

(iii)   Heisenberg Uncertainty

Heisenberg uncertainty principle states that without disturbing the system, pairs of quantum properties ' for example, position and momentum' cannot be exactly deliberated simultaneously; Thus it is impossible to calculate the quantum states of that system and the horizontal-vertical and diagonal polarization of photons are two for such pairs. The Heisenberg uncertainty depends on quantum theory that states a certain pair of physical properties are complementary. Thus, if one property is measured, the other is changed; so, any attempts to monitor the quantum state of a photon will change the properties and will be detectable and disturb the quantum system [10].

Hilbert space represents rectilinear basis in terms of the diagonal basis, considering the polarization of the photon in state of $| \updownarrow \rangle$:

$| \updownarrow \rangle = \frac{1}{\sqrt{2}} (|0\rangle + |1\rangle)$. This means that if the photon is sent vertically through a $45°$ polarizer, then the measured photon would be polarized with an angle of $45°$ with probability $(\frac{1}{\sqrt{2}})^2$ which equals $\frac{1}{2}$ [7].

(iv)   No Cloning Theorem

The exact single qubit quantum state can not be copied onto a different qubit state without disturbing the original qubit. This theorem was first shown by Wooters and Zurek in 1984. No cloning theorem prevents a third unauthorized party from decoding the transmitted information. Thus, if the eavesdropper tries to grasp the information, he/she can not do this without making a disturbance on the system. So, both the sender and the receiver detect the eavesdropping action. The photon polarization principle depends on this theorem, where unknown quantum state (qubit) can not be copied by an eavesdropper due to no-cloning theorem [2, 20].

(v)   Photon Entanglement

One of the key properties of quantum mechanical is the entanglement that is fundamentally different from the classical systems. Quantum entanglement has an important role in physical foundation and is an important resource in various quantum information processes. For example, it lets some physical process send two electrons in opposite directions. Measurement on each single electron has a certain probability of showing up or down spin. Nevertheless, the result of measurement on both electrons will always be in opposite spins. The universe allows the two electrons to communicate the outcomes of the first measurement instantaneously. Such an electron pair is called an entanglement and is in an entangled state which is a correlated state that is not a product state and it cannot be separated [21].

To explain entanglement, let's suppose a qubit $|\Psi\rangle$ in a zero $|0\rangle$ state.

Let U= H= $\frac{1}{\sqrt{2}} \begin{bmatrix} 1 & 1 \\ 1 & -1 \end{bmatrix}$, then let $|\Psi_1\rangle = H|\Psi\rangle = \frac{1}{\sqrt{2}}|0\rangle + \frac{1}{\sqrt{2}}|1\rangle$

Where:

U: Unitary Matrix of Hadmard space.

H: Hadmard Space.

$|\Psi\rangle$: Qubit state.

Now take another qubit $|\Psi_2\rangle$ with zero $|0\rangle$ state too, the joint state-space probability vector is the tensor product of these two states:

$|\Psi_1\rangle \otimes |\Psi_2\rangle = \frac{1}{2}|00\rangle + \frac{1}{2}|01\rangle + \frac{1}{2}|10\rangle + \frac{1}{2}|11\rangle$, so bits are entangled [6].

2.2.2 Quantum Logic Gates

A quantum logic gate is the analogue of a logic gate in a classical model. Just like we have AND, OR and NOT gates on a normal computer which take as input bits, a quantum computer has gates, which take as input quantum bits. Quantum logic gates can be modeled as matrix operations. Quantum computation are reversible computation, so evolution must be agreed with its reverse effect; operation with classical computation is represented by a truth table, but with quantum computation, operation is described by unitary transformation by using unitary matrix U which is an $2^n \times 2^n$.

Let $|\Psi\rangle$ be a state vector in n-dimensional Hilbert space (H_n), the operation of an n-qubit quantum logic gate can be represented by $|\Psi\rangle \rightarrow U|\Psi\rangle$ for some unitary $2^n \times 2^n$ matrix U. Quantum gates are often defined by simply specifying their matrices [3, 21].

a. One Bit Quantum Gates

The gates below are examples on this type of gates which are operating on single qubit. One qubit gates sample will be shown in Fig. (2.1.a) [6,19]:

1) Hadmard (H): Is denoted by H. This gate operates on a single qubit. When it operates on input $|0\rangle$, it outputs H ($|0\rangle$)=$\frac{1}{\sqrt{2}}$ ($|0\rangle + |1\rangle$). And for input $|1\rangle$, it outputs H ($|1\rangle$)=$\frac{1}{\sqrt{2}}$ ($|0\rangle - |1\rangle$).

$$H \rightarrow \frac{1}{\sqrt{2}} \begin{bmatrix} 1 & 1 \\ 1 & -1 \end{bmatrix}$$

2) Pauli_I: It is the identity gate which maps $|0\rangle$ to $|0\rangle$ and $|1\rangle$ to $|1\rangle$.

$$I \rightarrow \begin{bmatrix} 1 & 0 \\ 0 & 1 \end{bmatrix}$$

3) Pauli_X (NOT): NOT gate is the simplest gate which has a single input and single output; it just flips the input bit.

$$X \rightarrow \begin{bmatrix} 0 & 1 \\ 1 & 0 \end{bmatrix}$$

That is,

$$X|\psi\rangle = \beta|0\rangle + \alpha|1\rangle \tag{2.4}$$

4) Pauli_Y : This gate maps $|1\rangle$ to $-j\,|0\rangle$ and $|0\rangle$ to $j\,|1\rangle$.

$$Y \rightarrow \begin{bmatrix} 0 & -j \\ j & 0 \end{bmatrix}$$

5) Pauli_Z: The Z-gate inverts the phase of the qubit in $|1\rangle$ basis and leaves the basis state $|0\rangle$ unchanged.

$$Z \rightarrow \begin{bmatrix} 1 & 0 \\ 0 & -1 \end{bmatrix}$$

That is,

$$Z|\psi\rangle = \alpha|0\rangle - \beta|1\rangle \tag{2.5}$$

6) Phase_S: Phase gate S contains an angle of л/2 added to input state $|1\rangle$; thus, it maps $|0\rangle$ to $|0\rangle$ and $|1\rangle$ to $e^{i(\pi/2)}|1\rangle$.

$$S \rightarrow \begin{bmatrix} 1 & 0 \\ 0 & j \end{bmatrix}$$

b. Two-Bit Quantum Gates

Following are two-qubit transformation examples and their samples are shown in Fig. (2.1.b) [3, 19]:

1) CNOT: It also goes under the name (quantum) XOR. Its matrix is represented in the computational basis as:

$$CNOT \rightarrow \begin{bmatrix} 1 & 0 & 0 & 0 \\ 0 & 1 & 0 & 0 \\ 0 & 0 & 0 & 1 \\ 0 & 0 & 1 & 0 \end{bmatrix}$$

2) Swap: The unitary matrix is represented in this gate as:

$$SWAP \rightarrow \begin{bmatrix} 1 & 0 & 0 & 0 \\ 0 & 0 & 1 & 0 \\ 0 & 1 & 0 & 0 \\ 0 & 0 & 0 & 1 \end{bmatrix}$$

3) Controlled_ Z: The following matrix represents this gate basis:

$$Controlled\_Z \rightarrow \begin{bmatrix} 1 & 0 & 0 & 0 \\ 0 & 1 & 0 & 0 \\ 0 & 0 & 1 & 0 \\ 0 & 0 & 0 & -1 \end{bmatrix}$$

c. Three-Bit Quantum Gates

1) Toffoli gate: Any classical gate can be emulated by using this gate and this is performs an important property. Fig. (2.1.c) shows the three qubits quantum gates samples. The properties of this gate can be summarized as follows: [3, 6, 19]

$$T_F(a, b, c) = (a, b, c \text{ XOR}(a \text{ AND } b))$$

(2.6)

$$T_F(1, 1, x) : c' = \text{NOT } x$$

(2.7)

$$T_F(x, y, 1) : c' = x \text{ NAND } y \tag{2.8}$$

$$T_F(x, y, 0) : c' = x \text{ AND } y \tag{2.9}$$

$$T_F(x, 1, 0) : c' = a = a' = \text{FANOUT} \tag{2.10}$$

$$\text{Toffoli} \rightarrow \begin{bmatrix} 1 & 0 & 0 & 0 & 0 & 0 & 0 & 0 \\ 0 & 1 & 0 & 0 & 0 & 0 & 0 & 0 \\ 0 & 0 & 1 & 0 & 0 & 0 & 0 & 0 \\ 0 & 0 & 0 & 1 & 0 & 0 & 0 & 0 \\ 0 & 0 & 0 & 0 & 1 & 0 & 0 & 0 \\ 0 & 0 & 0 & 0 & 0 & 1 & 0 & 0 \\ 0 & 0 & 0 & 0 & 0 & 0 & 0 & 1 \\ 0 & 0 & 0 & 0 & 0 & 0 & 1 & 0 \end{bmatrix}$$

2) Fredkin (controlled-swap): In this gate, if the control line is set, it flips the second and third bits. The properties of the Fredkin Gate can be summarized as follows:

$$F_R(x, 0, y) : b' = x \text{ AND } y \tag{2.11}$$

$$F_R(1, x, y) : b' = c \text{ AND } c' = b, \text{ which is CROSSOVER} \tag{2.12}$$

$$F_R(x, 1, 0) : b' = a' = c = \text{FANOUT}, \text{ with } b' = \text{NOT } x \tag{2.13}$$

$$\text{Fredkin} \rightarrow \begin{bmatrix} 1 & 0 & 0 & 0 & 0 & 0 & 0 & 0 \\ 0 & 1 & 0 & 0 & 0 & 0 & 0 & 0 \\ 0 & 0 & 1 & 0 & 0 & 0 & 0 & 0 \\ 0 & 0 & 0 & 1 & 0 & 0 & 0 & 0 \\ 0 & 0 & 0 & 0 & 1 & 0 & 0 & 0 \\ 0 & 0 & 0 & 0 & 0 & 0 & 1 & 0 \\ 0 & 0 & 0 & 0 & 0 & 1 & 0 & 0 \\ 0 & 0 & 0 & 0 & 0 & 0 & 0 & 1 \end{bmatrix}$$

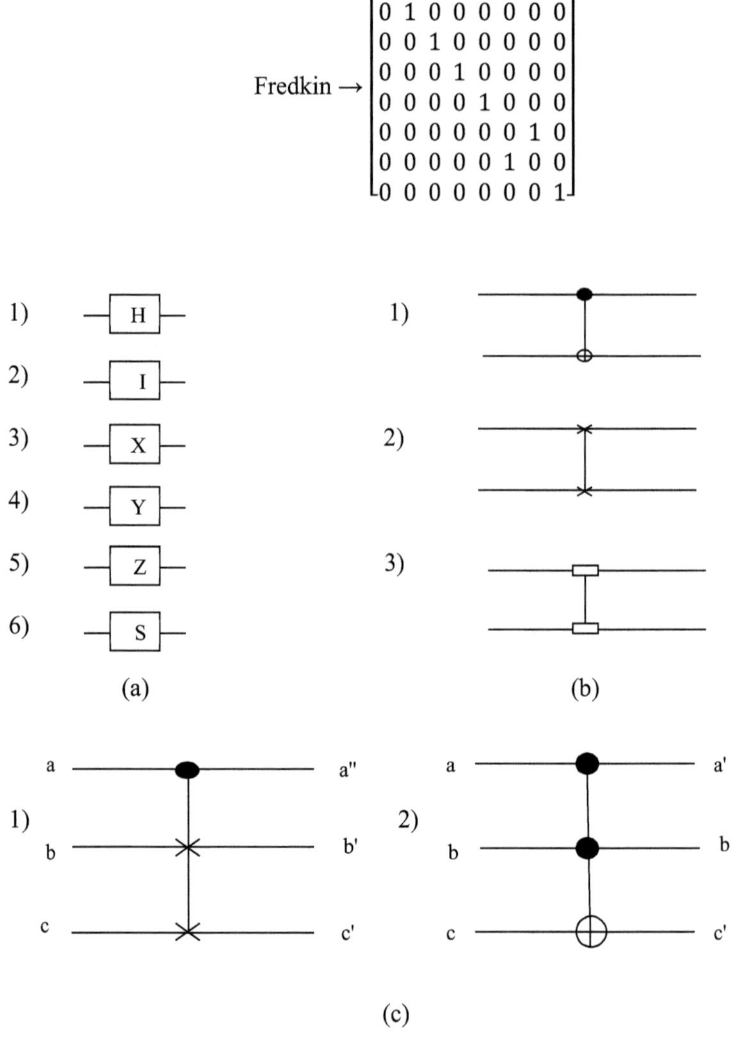

Fig. (2.1): (a) One qubit quantum gates, (b) Two qubits quantum gates, (c) Three qubits quantum gates.

2.3 Quantum Circuits

Any quantum circuit is a composition of parallel and serial connections of quantum logic gates. By using quantum logic gates, any quantum circuit could be built, and many operations in classical computation are realized and operated with quantum, using logic gates connected by wires carrying qubits, without fan out or feedback. Each line in quantum circuit represents a wire, and this wire does not necessarily corresponds to a physical wire; it may correspond instead to the passage of time or maybe to a physical particle such as a photon moving from location to another through space [12].

Quantum circuit diagrams are different from the classical diagrams in which they have the following constraints [18, 22]:

1. They are acyclic (no loops).

2. Quantum systems are reversible by nature. Information can travel freely in both directions: from inputs to outputs and vise versa. Thus, quantum gates have no FANIN, as FANIN implies that the circuit is not reversible, and therefore not unitary.

3. No FANOUT: One cannot copy a qubit's state during the computational phase because of the no-cloning theorem which prevents having:

$$|\Psi\rangle \rightarrow |\Psi\rangle|\Psi\rangle|\Psi\rangle \qquad \qquad ...(2.14)$$

All the constraints above can be simulated with the use of ancilla and carbage bits, if it is assumes that no qubits will be in a superposition as shown in Fig. (2.2). Ancilla bits are extra qubits needed for temporary calculations and the garbage bits are useless qubits left after computation.

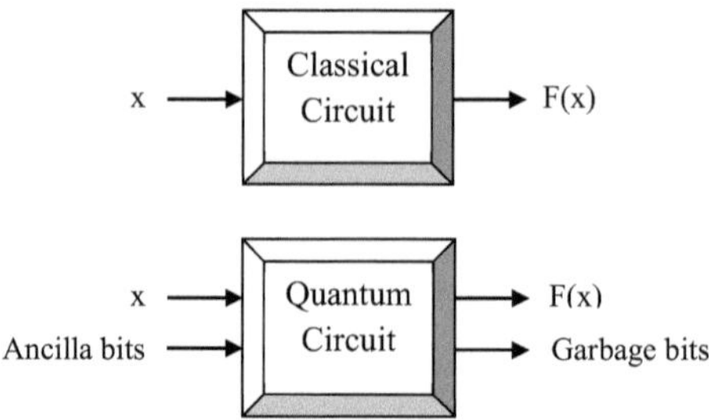

Fig. (2.2): Garbage and ancilla bits.

2.4 Field Programmable Gate Array (FPGA)

FPGA is a silicon chip containing an array of configurable logic blocks (CLBs), which are connected together through a matrix of vast interconnection to form complex digital circuits. Fig. (2.3) shows the FPGA architecture. The Configuration Logic Blocks (CLBs) look like islands bounded by a sea of interconnection. Each CLB contains a Lookup Table (LUT) which is configured to implement a prehistoric logic gate. The interconnection is also configurable, and it connects CLBs together so it can compose more complicated circuits from the primitive logic gates. The advantage of FPGAs over the traditional microprocessors is that FPGAs still benefit from a clock-speed increase with each new generation of hardware [24, 25].

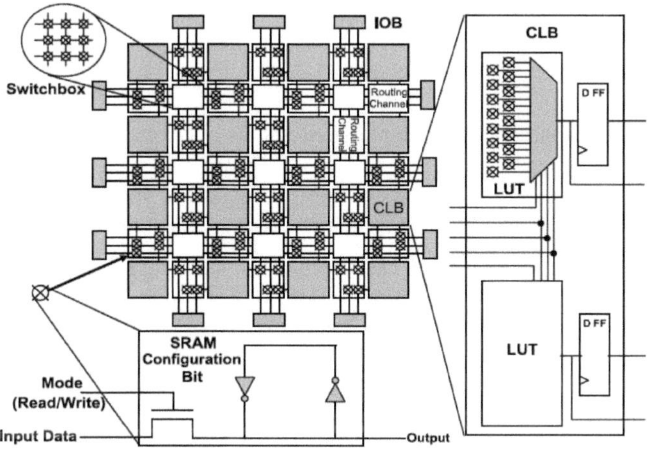

Fig. (2.3): Architecture of a typical FPGA.

2.5 Programming Languages

Manual methods are feasible for designing logic circuits only when the circuits are small. For practical circuits or anything else, computer-based design tools are used by designers. Computer-based design tools and correct-by-construction methodology leverage the creativity and effort of a designer and reduce the risk of producing a flawed design; prototype integrated circuits are too expensive and consume time to build. Thus all modern design tools rely on a Hardware Description Language (HDL) to describe, design, and test a circuit in software before it is ever manufactured [26, 27].

2.5.1 Hardware Description Language

A hardware description language (HDL) is a computer-based language that describes the hardware of digital systems in a textual form. FPGA designs are programmed in either VHDL ((Very high speed integrated circuit) Hardware Description Language) or Verilog in the process of being compiled into the design which is totally transparent to the user and does not require the user to know anything specific about either of these two Hardware Description Languages (HDLs). There are some HDL languages such as VHDL, Verilog®-HDL, Verilog®-A, VHDL-AMS, Verilog®-AMS [25].

1. <u>VHDL</u>:-

VHDL is a high-level hardware description language used to describe digital circuits that can be programmed into an FPGA. The initial declaration of VHDL design is an entity declaration and an architecture body. The entity declaration describes the design I/O and includes parameters that customize the entity. The architecture body describes the internal working of the entity. VHDL is not case sensitive and does not require a special formatting, such as spaces, tabs, or indentation. Each line of code statement must end with a semicolon. Filename extension can be either .vhd or .vhdl [26, 28].

2. <u>Verilog HDL</u>:-

Verilog HDL (sometimes referred to as Verilog®-D for digital) is a hardware description language (HDL) used to model electronic systems. It is a programming language that has been designed and optimized for describing the behavior of digital systems. Verilog HDL language is case sensitive, meaning upper and lower case letters are different. Spaces are important [29].

3. Verilog®-A:-

Verilog®-A is developed due to the need to model analogue circuit behavior; an analogue-only specification provides a unique set of features over the digital modeling language [29].

4. VHDL-AMS:-

For mixed-signal (analogue and digital), two modeling languages are emerging; electronic and mixed-technology system modeling; Verilog®-AMS and VHDL-AMS. Both extensions from the digital domain are generally referred to as Analogue and Mixed-Signal (AMS) languages for electronic circuits. This manner leads to the modeling of nonelectrical and electronic parts using the same model constructs [29].

2.5.2 How to Select FPGA Type for any System

There is a wide range of FPGAs provided by many semiconductor vendors including Xilinx, Altera, Atmel, and Lattice. Each manufacturer provides each own unique architecture [17]. Common architecture of the FPGA is the same among different FPGA manufactures. FPGA consists of a group of programmable gates connected together with flexible interconnections. These gates are implemented using lookup tables (LUTs) for computational units, flip-flops or timing, switchable interconnected for routing, and input-output blocks (IOBs) for transferring data into and out of the device [26].

The configurable of the FPGA may be stored in many numbers of ways and a number of programming technologies and their differences have a significant effect on programmable logic architecture. The approaches that have been used historically include EPROM (Erasable Programmable Read Only Memory), EEPROM (Electrically Erasable Programmable Read-Only Memory), flash, static memory, and anti-fuses. Of these approaches, only the flash, static memory and anti-fuse approaches are widely used in modern FPGAs [26].

The SRAM makes the FPGA volatile; i.e., it must be programmed every time it is started up. The SRAM programming bits are distributed across the entire FPGA, and stored locally with the LUTs that are switchable interconnects. Data is stored or read from the SRAM cells as long as the cell is powered; without power, the SRAM cell lost its value [26].

Flash memory is a high-quality programmable read-only memory. Flash uses a floating gate structure in which a low-leakage capacitor holds a voltage that controls a transistor gate. This memory cell can be used to control programming transistors. The memory cell controls two transistors. One is the programmable connection point. It can be used for interconnecting electrical nodes in interconnect or logic. The other allows read-write access to the cell [29].

In this thesis, Altera DE2-115 cyclone IV E is used with flash memory technique. The distribution system is programmed using VHDL programming language. Because Altera FPGA kit is available at the lab and flash memory technique is used since the data which produced and processed inside the FPGA is transferred immediately after the processing and the distribution is done.

Chapter Three

Quantum Cryptography

3.1 Theoretical Background

Private key distribution used the same key in both encryption and decryption of the message, which is a required beforehand secure key. Thus, both sender and receiver should meet to handover the secret key to each other; public key distribution secrecy can be cracked through renders messages insecure retroactively. Thus, Quantum cryptography was developed to solve the key distribution problems with the classical cryptographic techniques mentioned before.

3.2 Quantum Cryptography

Quantum cryptography is a technique to distribute a sequence of truly random and unconditionally secure bits over a secure communication by applying the phenomena of quantum physics. Its goal is to provide a secret key which is truly random and unconditionally secure, as long as the message between two authorized parities who never met before and not need to meet. The sharing process of the unconditionally secure key is done in the presence of an eavesdropper since quantum cryptography security is guaranteed by the principles of quantum mechanics which make the eavesdropper unable to catch the right copy of the unknown quantum state [30].

Quantum cryptography is only used for key distributing but not for transmitting any useful data. Thus, it is called Quantum Key Distribution (QKD) [12].

Quantum cryptography security mainly depends on two quantum mechanics: the principle of photon polarization and the Heisenberg Uncertainty principle. The photon polarization principle describes how the photons of light can be polarized in an exact direction which prevents the eavesdropper from truly measuring the quantum states as mentioned before in chapter two.

Heisenberg uncertainty principles which make the disturbance of the quantum system clear to the sender and the receiver when an attacking operation occurs to that system.

3.3 Quantum Cryptographic Protocols

There are many quantum key distribution protocols such as: BB84, B92, SARG04, E91, COW, DPS and S09 [7, 8, 12, 13]. The most important protocols will be discussed in brief.

3.3.1 Bennett-Brassard (BB84) Protocol

BB84 is the most widely used protocol nowadays. It is the first QKD protocol, developed by Bennett and Brassard in 1984 [2, 11].

Except for the BB84 protocol, the other protocols have the secure identification, handling weak pulse of detector and creating a bit error during the communication problems; because of these problems, BB84 is the best QKD protocol [16].

Devi and Raj [14] implemented a quantum cryptography system based on BB84 protocol and proved that BB84 protocol has a Quantum Bit Error Rate (QBER) lower than other quantum cryptography protocols.

The unconditional security of the BB84 was proved by Mayer's, who proved that BB84 protocol is an unconditionally secure from the attacker and proved the mathematical secrecy of this protocol [11].

In BB84 protocol coding scheme, four non-orthogonal polarization states (0°, 90°, 45° and -45°) are used to polarize each of the photons that will be sent as shown in Table (3.1). The communication in this protocol requires two channels: quantum channel (e.g.: optic fiber or free space) and public channel (e.g.: Internet). Alice and Bob have to communicate within channels, quantum channel and public channel to share a secret key. First, Alice and Bob have to communicate (one way communication) via quantum channel, and then they both will create a public connection over the public channel (two way communication) [2,5].

Table (3.1): Polarization of quantum bits.

State	Basis	Value	Polar angle
$|0\rangle$	+	0	0°
$|1\rangle$	+	1	90°
$|0\rangle+|1\rangle$	x	0	45°
$|0\rangle-|1\rangle$	x	1	-45°

In the rectilinear (+) base, the photon is sent in either horizontal base with 0° which is 0 or in vertical base with 90°; then it performs 1. In the diagonal (×) base, the photon is sent with -45° to define 1 or with 45° to define 0 [20].

a - BB84 Protocol Operations over Quantum Channel

Stage1: Alice will send polarized photons (measured as bit) to Bob using the Quantum channel. For example if Alice wants to send 11010110 to bob Alice will choose randomly basis (either the rectilinear or the diagonal) as shown in Fig. (3.1); then the photons are polarized according to the bases which has already been described [21, 31].

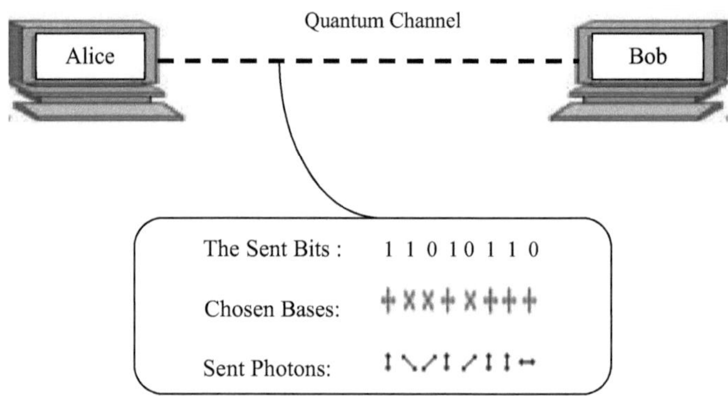

Fig. (3.1): Alice's polarized photons transmitted over quantum channel.

Stage2: After all the photons have been transmitted, Bob will measure the bits which have been received by using the rectilinear or diagonal basis as shown in Fig. (3.2). If Bob chooses the same basis chosen by Alice, a correct state which is a correct quantum bit will be received; it is called raw key or sifted key [21, 31].

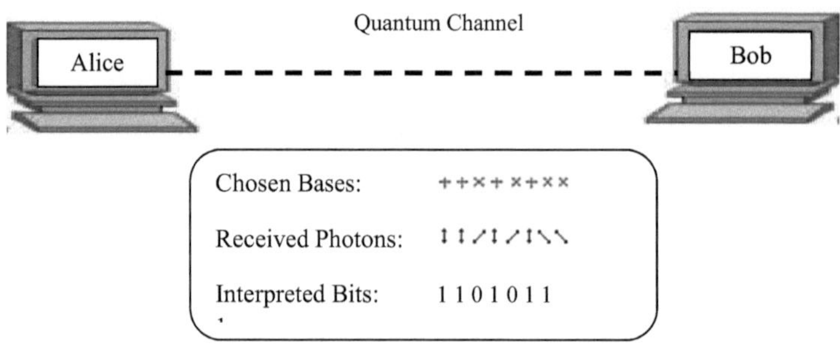

Fig. (3.2): Bob's measurements.

b - BB84 Protocol Operations over Public Channel

Both Alice and Bob will establish a connection to communicate over a public channel. The stages done over the public channel are: sifted key stage, error correction stage and privacy amplification [20]. Fig. (3.3) shows the public channel stages.

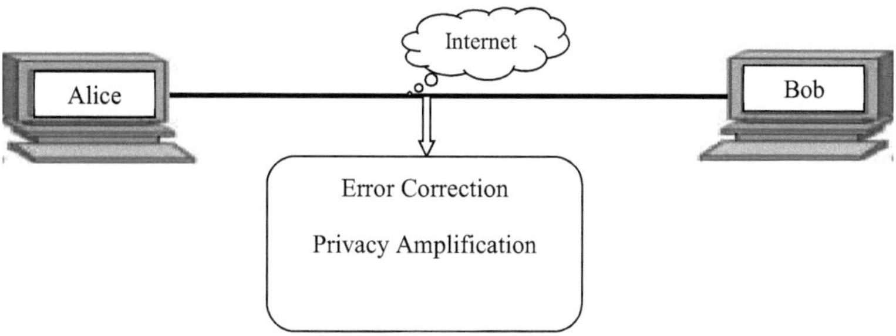

Fig. (3.3): Public channel operations.

(i) Sifting Key Stage

In this stage which is shown in Fig. (3.4), Alice and Bob negotiate which bits are used and which bits are discarded. Bob will proclaim his measurements (states) at the public channel with or without the presence of eavesdropper. Alice will retort the correct measurements that Bob measured with or without the presence of an eavesdropper. Now, Alice and Bob share a raw key, which is considered not fully a secret; bits may be tampered by eavesdropper during the transmission [9, 31]. Both Alice and Bob discard bits number 2,7 and 8 (numbers from left to right), so, the sifted key will be 10101.

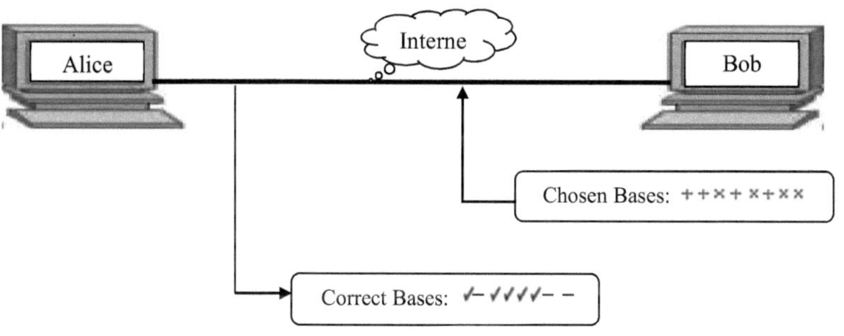

Fig. (3.4): Sifting key stage.

(ii)   Error Correction Stage

Error correction stage splits into two sub stages: first, the error estimation which estimates the error rate in the sifted key or raw key; second, error reconciliation sub stage which cancels the error bits and correct them.

Error estimation is an important step in the QKD protocol to calculate the appropriate error rate in the sifted key. The BB84 error rate p is estimated by selecting a small random subset of bits with length r from those given in the sifted key.

This test string is done publicly over the quantum channel by Alice and Bob and yields in a certain number of errors e. If the length of the test string chosen is sufficient to the length of the sifted key n, then, the probability of error will be [20]:

$$p = \frac{e}{r} \tag{3.1}$$

In BB84 protocol the maximum error probability is 25%, because, if Alice sends quantum bit with logic 0 in rectilinear base (0, state →), and Bob chooses the rectilinear base with → state to decode this polarized bit or photon, then, he will receive a correct state since he used the same base as the one Alice chose.

But, if Bob chooses the diagonal base to decode it, base ×, the wrong base then, he will receive either $|/\rangle$ state or $|\backslash\rangle$ state with equal probability. Thus, the probability that Bob will receive the correct state is 75% [24].

The random bits used for error estimation are then discarded from the raw key, since the error will occur because of the quantum channel noise which makes the 1 received by Bob as 0 and vise versa and so, after the error estimation, Alice and Bob correct the noisy bits (error bits) [20].

The error will be corrected by some algorithms like CASCADE algorithm. CASCADE appropriate part for the error correction is BINARY as shown in Fig. (3.5). When a number of errors in Bob's string is an odd number, both Alice and Bob perform an interactive binary search on the strings to find and correct one error, respectively. At the beginning, Alice sends to Bob the parity of the entire string and Bob checks if his parity bit differs from Alice's parity bit; if both bits are the same, there is an even number of errors (possibly zero) in the string and nothing is done. If the parity bits are different, the error will be found by applying the following steps: [20]

1. Alice sends to Bob the parity bit of the first half of the string.

2. Bob determines if there is an odd number of errors in the first or in the second half by testing the parity of his string and compare it to the Alice's parity.

3. With the half determined in step 2, the operation starts again at step 1, until the erroneous bit is found.

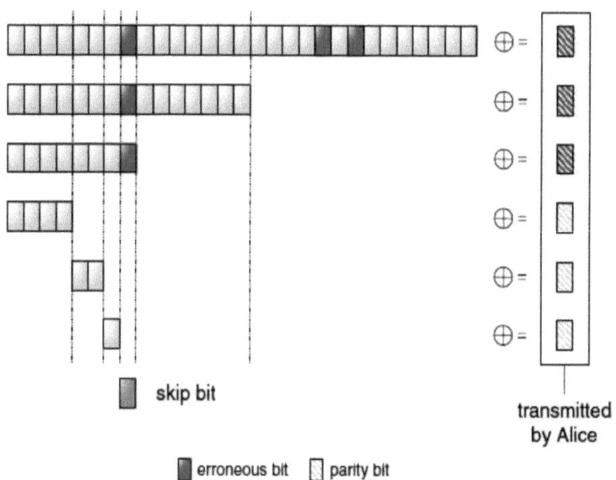

Fig. (3.5): BINARY corrects exact one error of a block with an odd number of errors. [20]

(iii) **Privacy Amplification Stage**

Privacy amplification is the last stage of the public channel processes and BB84 protocol. This stage is done by the extraction of final secret key by randomly choosing a subset of bits from the reconciled key. This is done by using the information on error rate and the upper bound of the number of the bits that Eve could poses, since the reconciled key is partially secure [31].

The reconciled key can not be used as a secret key between Alice and Bob because eavesdropper information about the reconciled key must be considered since he or she may gain information during the error correction and also maybe during the transmission via the quantum channel. Thus, Alice and Bob must map their string via a function to a smaller subset, so that eavesdropper's knowledge decreases almost to zero. After this stage which is the privacy amplification stage, Alice and Bob share a secret key only known by them [20].

In [20], a good example of privacy amplification supposed that Bob sent his random bases to Alice and it was 0010011101011000101; after the reconciled stage, the raw key between Alice and Bob is 1101001110. Due to the error rate and the information leaked during reconciliation, Alice and Bob computed that Eve has 4 bits of the information, so the key length must be shorten to 6 bits. The value of c (where c is a value chosen randomly by applying a hash function or Alice and Bob could use a random string from the reconciled key as c) is extracted from Bob's random choice of bases, which is 0010011101 and multiply c with the reconciliation key.

$$0010011101 \cdot 1101001110 = 10000001101101010110$$

The final key consists now of the last six bits of the result (010110).

3.3.2 SARG04 Protocol

SARG04 is a modified version of the BB84 protocol. This protocol makes quantum key distribution robust against photon-number-splitting attacks. SARG04 protocol uses the same four states as the one in BB84 exactly; SARG04 differs from BB84 in the encoding and decoding of classical information. In this protocol, Alice announces one of the four following sets publicly:

$\{|\Psi_1\rangle, |\Psi_3\rangle\}, \{|\Psi_2\rangle, |\Psi_3\rangle\}, \{|\Psi_1\rangle, |\Psi_4\rangle\}, \{|\Psi_2\rangle, |\Psi_4\rangle\}$

The SARG04 protocol goes as follows:

1. Alice prepares m qubits randomly for each qubit in one of the four states $|\Psi_1\rangle$, $|\Psi_2\rangle$, $|\Psi_3\rangle$ or $|\Psi_4\rangle$ and sends them to Bob over a quantum channel.

2. For each qubit received by Bob, Bob chooses randomly one of the two bases: $\{+, \times\}$.

3. For each qubit Alice sent, Alice announces publicly one of the four sets $\{|\Psi_1\rangle, |\Psi_3\rangle\}, \{|\Psi_2\rangle, |\Psi_3\rangle\}, \{|\Psi_1\rangle, |\Psi_4\rangle\}, \{|\Psi_2\rangle, |\Psi_4\rangle\}$, that contains the photon states sent out by her.

4. Bob tells Alice to discard the times when the measurement output is confusing.

5. Then, Alice and Bob test their key security by using a randomly chosen key and they compare the results of their subset. If errors are detected, they abort and start again because their transmission is insecure.

6. To generate a secure key, techniques of error correction and privacy amplification are used.

7. To encrypt a message, the one time pad is used.

Table (3.2) shows an example of SARG04 protocol from Alice to Bob.

In the first column of Table (3.2), Bob can not determine the sent state because Ψ_4 state is measured in the basis $\times$ that must be 0 and Ψ_1 state is measured in the basis $\times$ that could be 0; therefore, Bob can not determine the sent state whether it is Ψ_1 or Ψ_4. The second column of the same table shows that Bob knows for sure that the sent state can not be Ψ_4, which, all the time, gives a measurement of 0 if measured in the basis $\times$ [8]

Table (3.2): An example of SARG04 protocol from Alice to Bob. [8]

The Bit	1	2	3	4	5	6	7	8	9	10	11	12	13	14	15
Alice's Random Bits	0	0	0	1	0	1	1	1	0	1	0	1	1	0	1
Alice's States	Ψ_1	Ψ_1	Ψ_1	Ψ_4	Ψ_3	Ψ_4	Ψ_2	Ψ_2	Ψ_3	Ψ_4	Ψ_1	Ψ_2	Ψ_4	Ψ_3	Ψ_2
Bob's Random Basis	×	×	+	×	×	+	+	×	×	×	+	×	+	+	+
Bob's Results	0	1	0	1	0	1	1	1	0	1	0	1	1	1	1
Announcement	Ψ_1	Ψ_1	Ψ_1	Ψ_1	Ψ_2	Ψ_2	Ψ_2	Ψ_2	Ψ_1	Ψ_1	Ψ_1	Ψ_2	Ψ_2	Ψ_2	Ψ_2
States	Ψ_4	Ψ_4	Ψ_3	Ψ_4	Ψ_3	Ψ_4	Ψ_3	Ψ_4	Ψ_3	Ψ_4	Ψ_3	Ψ_3	Ψ_4	Ψ_3	Ψ_3
Discovered States		Ψ_1				Ψ_4						Ψ_2	Ψ_4	Ψ_3	
The Sifted Key		0				1						1	1	0	

3.3.3 B92 Protocol

B92 is a two- state quantum key distribution protocol. It is secure in theory but from the practical point of view, this protocol has low level security because the encoded bits can simply decoded since it encodes the data with two states only. B92 states are:

$$|\Psi\rangle = \begin{cases} |0\rangle & \text{if the bit is 0} \\[2em] \dfrac{|0\rangle + |1\rangle}{\sqrt{2}} & \text{if the bit is 1} \end{cases}$$

The communication steps between Alice and Bob can be followed by the following steps:

1. Alice generates a random binary sequence as she did in the BB84 protocol and polarized the quantum bits as photons with the two above states and sends them to Bob over the quantum channel; Alice sends the 1 as ↗ and the 0 as ↔.

2. Bob receives this sequence of qubits and generates a random sequence independent of Alice's random sequence since Bob has no idea about Alice's random sequence.

3. Bob saves measurement after communicating with Alice over the public channel. Alice and Bob put 1 in the position when their bases are different and put 0 at the same bases positions. [2, 30]

3.4 Eavesdropping on Quantum Key Distribution

QKD security is a fine theoretical result, but this does not mean that real world QKD systems would be secure. QKD could be eavesdropped; eavesdropping is the art of intercepting and reading messages and conversations by unauthorized recipients.

The eavesdropper is someone who listens secretly to the conversations of others. Eavesdropping can be done over telephone lines, emails, instant messages, and any other method of private communication [30].

3.4.1 Intercept/ Resend Attack

In this type of eavesdropping or attacking type, an eavesdropper intercepts or taps pulse from the sender and reads it in his/her bases. The eavesdropper (Eve) measures each pulse as qubits in one of the quantum bases as Bob does; then, Eve will pretend as Alice and resends another qubits to Bob corresponding to her measurement results. Eve and Bob have the same chance to measure the qubits sent by Alice since, both Eve and Bob do not know the random number generator used by Alice. To avoid this type of attacking, many techniques could be used; one of these techniques is to use double QKD protocol layers [9, 11].

3.4.2 Beam Splitting Attack

It is so difficult to prepare precise calculation of one-photon states in practical quantum cryptography. Eve could gain one photon and split its state to several states and take the original one and forward the other to Bob; Eve can have the opportunities to split the signal and learn partial information on the key without making a disturbance on the transmission between Alice and Bob. To avoid this type of attacking, quantum protection devices are needed [11, 12].

Chapter Four

Design and Implementation of Quantum Key Distribution Using FPGA

4.1 Theoretical Background

It has already been discussed that there are different QKD protocols. In this thesis, BB84 protocol is used to implement quantum key distribution system which is composed of two stages, each with different base probability value. This system will increase the protocol gain without decreasing the security level in both sifting key stage and error elimination stage.

The generation components are simulated using Linear Feedback Shift Register (LFSR) over the Field Programmable Gate Array (FPGA); after that the transmitting of the key bits is simulated over the computer using MATLAB R2010a.

4.2 The Proposed Model

The encryption sustem which presents the transceiver of Quantum Key Distribution (QKD) system is shown in Fig. (4.1). The proposed system deals with sharing a secure key over the quantum communication channel.

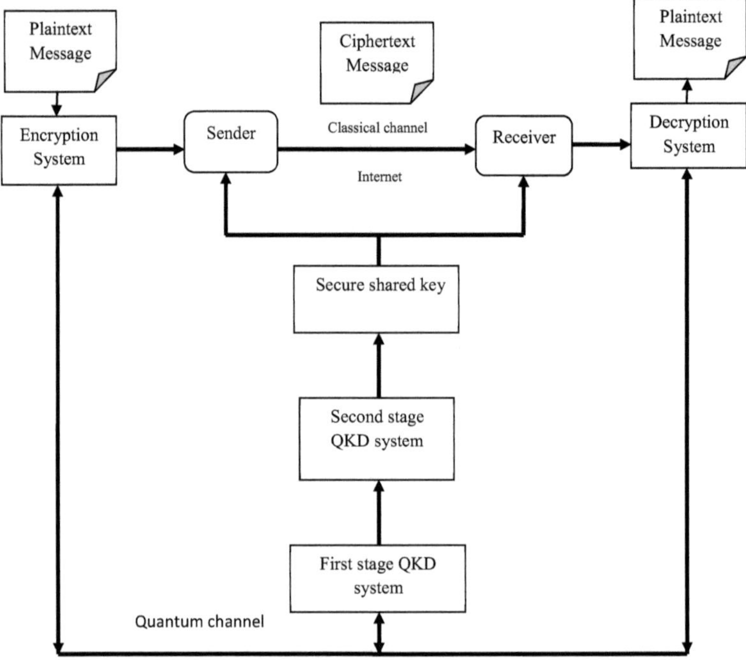

Fig. (4.1): Encryption system structure.

4.2.1 First Stage QKD System

Fig. (4.2) shows the flowchart of the first stage of the QKD proposed system. In this model, different values of base probability and different kinds of gates inputs are taken to enhance the system performance.

The simulation parameters used in this model are shown in Table (4.1). A comparison is made among six cases; each case has a different simulation parameter in order to show the best case among these cases.

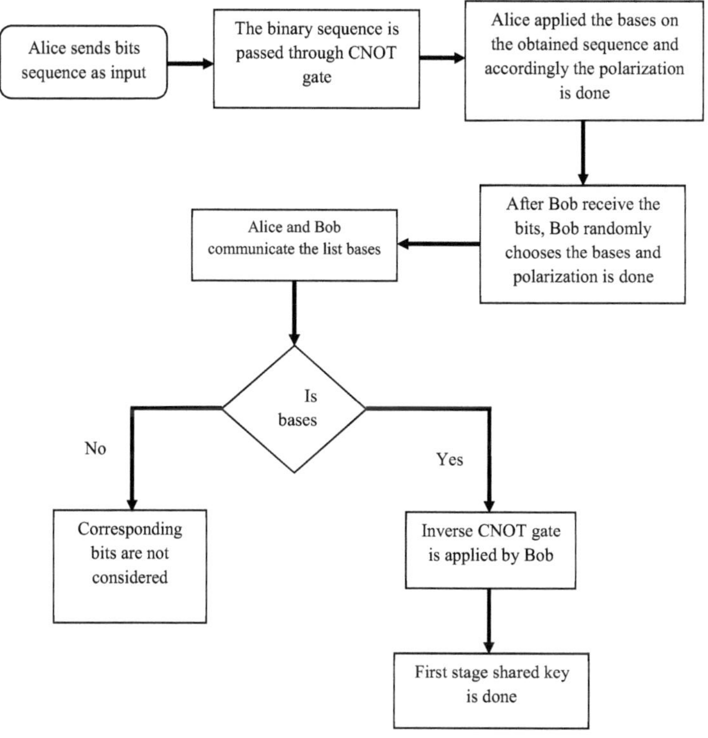

Fig. (4.2): First stage QKD flowchart.

4.2.2 Second Stage QKD System

Second stage QKD system is shown in Fig. (4.3). In this stage, all the system cases have the same base probability which is 50%, but the CNOT gate bits are the same as in the first QKD stage.

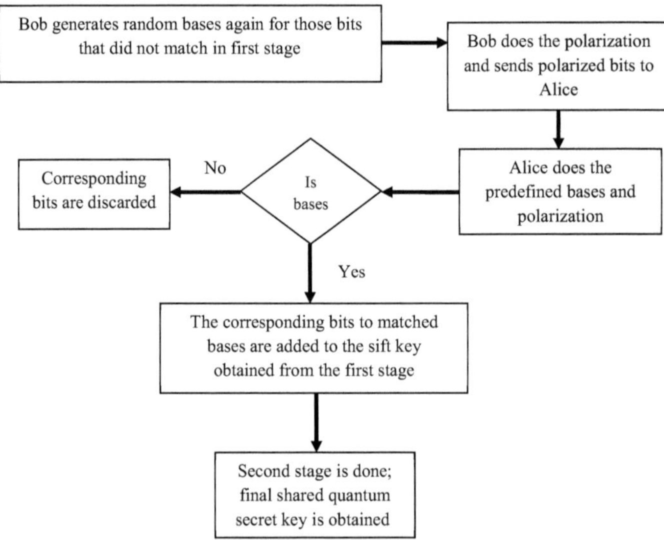

Fig. (4.3): Second stage QKD flowchart.

Table (4.1): Simulation parameter of the QKD proposed system.

Number of QKD Stages	Two Stages BB84 Protocol
Coding Method	CNOT Quantum Gate
1st Stage Base Probability	60%
2nd Stage Base Probability	50%
Qubits Geneorator Algorithm	LFSR
Bases Generator Algorithm	LFSR
Qubits and Bases Lengths	256, 512, 1024, 2048, and 4096 bits
Bases Simulation Coding	$0 \rightarrow$ Rectilinear Bases $1 \rightarrow$ Diagonal Bases

The proposed quantum key distribution is based on three factors:

1. Number of repeating the QKD protocol.
2. Base probability.
3. N-bits C-QUBITS technique.

Most of the recent papers combine the BB84 protocol and B92 protocols or combine B92 protocol twice for improving the efficiency and performance.

Base probability δ performs the polarization base distribution in the base space. When δ is 0.5, the probability to send a qubit in the rectilinear base or diagonal base is 50% but when δ is 0.6, the probability to send a qubit in the rectilinear base or diagonal base is 60% and so on. The protocol gain is increased whenever the base probability increased as mentioned in [20]; when δ is 0.5, the protocol gain is in its worst case and the enhancement starts from $\delta=0.6$. The fact that δ is variable gives us the probability to increase the protocol gain according to Eq. (4.1).

$$g_p = \delta \cdot \delta + (1 - \delta) \cdot (1 - \delta) = 2\delta^2 - 2\delta + 1 \qquad \ldots(4.1)$$

Where g_p is protocol gain. Fig. (4.4) shows a graph to the base probability against the protocol gain.

It should be mentioned that increasing the value of δ leads to decreasing the security level as explained in [20].

Adding quantum gate to the bits before polarization and sending decreases the maximum probability of bits that could be catched by Eve. Adding 2-bits quantum gate gives four possible quantum states to the bits $|00\rangle, |01\rangle, |10\rangle,$ and $|11\rangle$, so Eve can catch maximum $\frac{1}{4}$ of the qubits the sender sends to the receiver. By using 3-bits quantum gate, qubits have eight different states $|000\rangle, |001\rangle, |010\rangle, |011\rangle, |100\rangle, |101\rangle, |110\rangle,$ and $|111\rangle$ and this gives Eve maximum bits to catch equal to $\frac{1}{8}$ of the qubits sent by the sender to the receiver and so on.

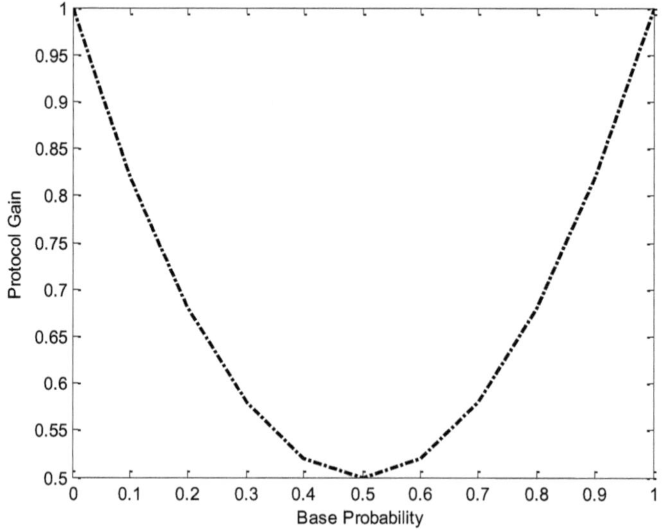

Fig. (4.4): Protocol gain g_p when delta δ is variable (Eq. 4.1).

As mentioned before, in this chapter, six cases were examined according to the previous factors; each case consists of two BB84 QKD protocol, generates the bases and the quantum bits presentation using the FPGA with LFSR algorithm and transmits them from the FPGA to the computer using serial (USB to RS232) cable. The transmitting starts later then sifting phase is started over UDP Ethernet protocol in Matlab language. The lookup table of the 2-bits CNOT gate is shown in Table (4.2); while Table (4.3) shows the lookup table of the 3-bits CNOT quantum gate.

Table (4.2): Truth table of 2-bits CNOT quantum gate.

Input		output		Type
A (control)	B (XOR)	A	B	
0	0	0	0	Identity
0	1	0	1	Identity
1	0	1	1	Swap
1	1	1	0	Swap

Table (4.3): Truth table of 3-bits CNOT quantum gate.

Inputs			Outputs		
A	**B**	**C**	**A'**	**B'**	**C'**
0	0	0	0	0	0
0	0	1	0	0	1
0	1	0	0	1	0
0	1	1	0	1	1
1	0	0	1	0	0
1	0	1	1	0	1
1	1	0	1	1	1
1	1	1	1	1	0

Using two stages QKD protocol increases the sifted key length by giving second chance to the receiver to correct its independent bases; the wrong base chosen by the receiver will be decreased and the number of compatible bases is increased, so the discarded bits will be minimum. Using two steps and base probability with 0.6 will increase the protocol gain but it will weaken the security level of the system due to the possible leakage in information that should be exchanged between the sender and the receiver because the overall system information (i.e. the number of steps and base probability) should be exchanged between sender and receiver before the key distribution operation started; hence, by adding the CNOT gate, the chance of Eve to encode the catched bits will be decreased, so the system security level will be kept.

4.3 Implementation of the Proposed Model Using FPGA

The proposed model of the system implementation using FPGA is shown in Fig. (4.5) below. The binary sequence bits generated by the sender in order to start the QKD protocol are generated by applying the LFSR algorithm using the FPGA kit.

The sender should enter the length of the binary sequence and the initial state value of the LFSR shift register to the FPGA internally using Quartus II web edition of the ALTERA DE2-115, which contains Altera Cyclone® IV 4CE115 FPGA device.

LFSR Algorithm is used to generate the polarity bases in which logic '0' performs rectilinear polarity bases and logic '1' is the diagonal polarity bases. Bases generator is implemented with Quartus II web edition over the FPGA kit with base probability 60% to the first stage and 50% to the second stage.

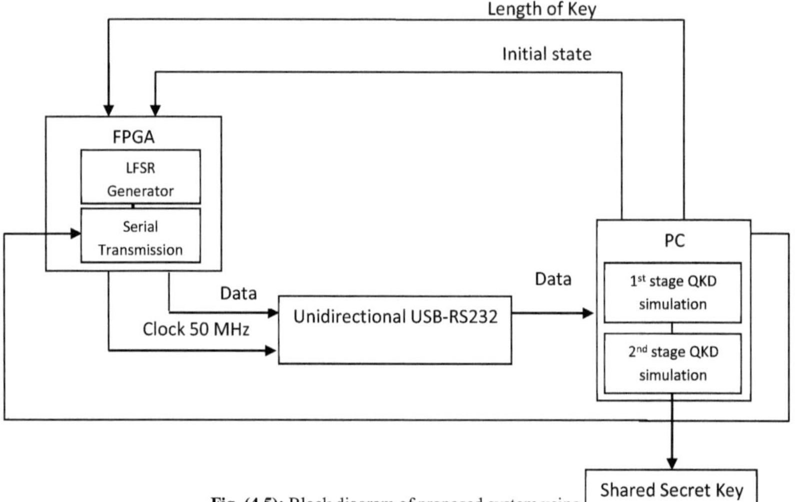

Fig. (4.5): Block diagram of proposed system using

4.3.1 LFSR Binary Bits Generator Using FPGA

The "pseudo-random" number generator (PRNG) is used to generate binary bits; Linear Feedback Shift Register (LFSR) algorithm is applied to generate the bits.

LFSR is a feedback shift register which made up of two parts: Shift register and Feedback function. The feedback function is simply the XOR of certain bits in the register; the list of these bits is called a **tap sequence.** A LFSR with n flip-flops can implement only a (2^n-1) state counter. The all-zeros state is normally not allowed because the counter locks up. Good design practice demands a reset condition that provides startup in a known condition and also ensures that the counter does not power up in a zero condition and stay locked up. The choice of the polynomial used should ensure 2^n-1 states with no repeated states; such a polynomial is known as a primitive or maximal-length polynomial.

The following figure, Fig. (4.6) shows a typical LFSR bit generator.

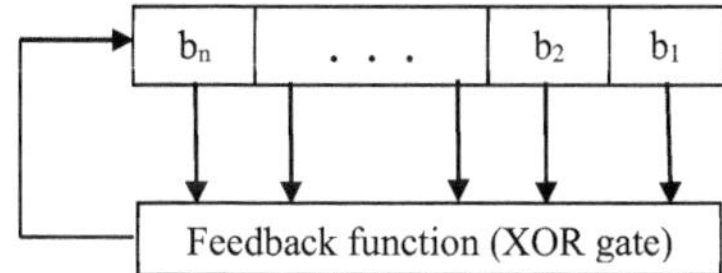

Fig. (4.6): Typical linear sequence generator using LFSR.

In the proposed Model, LFSR is used to generate 256, 512, 1024, 2048, and 4096 bits as the plaintext message length, Table (4.4) shows the locations of the feedback of the proposed model LFSRs generators.

Table (4.4): Taps for Maximal-Length LFSRs with 8 to11 Bits.

No. of bits	Polynomial Equation	Period 2^n-1
8	$X^7+X^3+X^2+X$	255
9	X^8+X^3	511
10	X^9+X^2	1023
11	$X^{10}+X^1$	2047
12	$X^{11}+X^5+X^3+1$	4095

LFSR Algorithm is implemented over the FPGA kit, Altera DE2-115 which contains Altera Cyclone® IV 4CE115 FPGA device. The initial state bits of the LFSR shift register entered to the FPGA kit internally using Quartus II web edition with VHDL programming language.

The bits are generated after the end of the compilation and synthesis without waiting for a clock pulse because the algorithm is implemented as one state which speeds up the generation process within the FPGA clock period which is 20 ns. Fig. (4.7) shows a flow chart of the LFSR binary bits generator using FPGA.

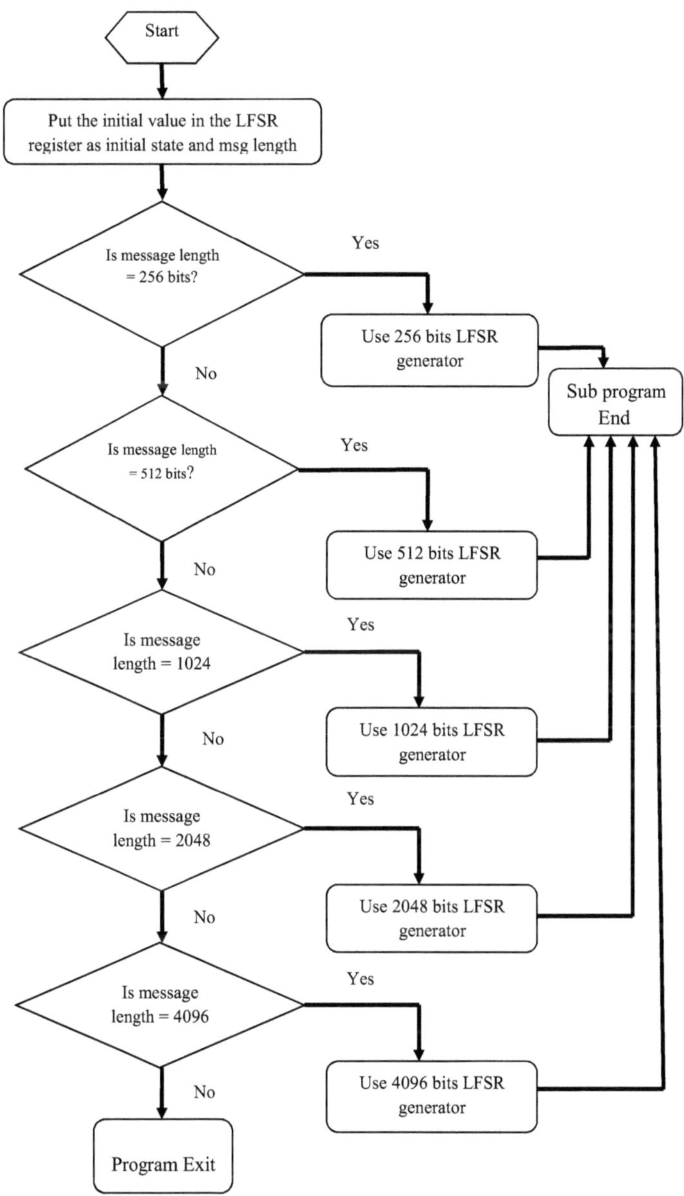

Fig. (4.7): LFSR based FPGA generator flowchart.

4.3.2 Base Generator Using FPGA

LFSR is used in proposed model as a base algorithm; BB84 bits are polarized with the bases generated from this algorithm. In this model, the bases are simulated in order to simulate the polarization bits; thus the bases can be presented and simulated as follows: 0 for horizontal base and 1 for vertical base. The proposed system simulate the bases: 0 for rectilinear and 1 for diagonal. Table (4.5) shows the bases and their binary equivalent value.

Table (4.5): Quantum bases and their binary simulation.

Rectilinear base +	↑	→
Diagonal base ×	╲	╱
Bit value	1	0

To simulate the bases as binary value, the LFSR algorithm is used to generate the binary presentation of bases; the same algorithm which is used to generate the binary bits will be used to generate the bases but with the required base probability. Fig. (4.8) shows the bases flowchart generator. Tabel (4.6) contains the FPGA simulation parameter of the LFSR algorithm.

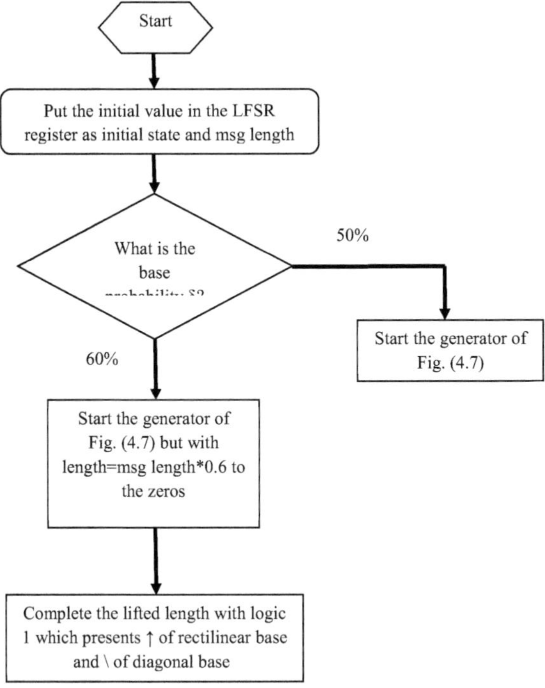

Fig. (4.8): Base generator flowchart.

Table (4.6): FPGA simulation parameter.

Number of Used Pins	Three Pins
Execution Clock Period	20 ns
Transmitting Media	Serial, RS232 USB Cable
Transmission Frequency	50 MHz
Transmission Baud rate	115200 bps

4.3.3 Implementing First Stage QKD Using FPGA

As mentioned before, first stage QKD is implementing with base probability 0.6; Fig. (4.9) below shows the block diagram of the implementation of this stage.

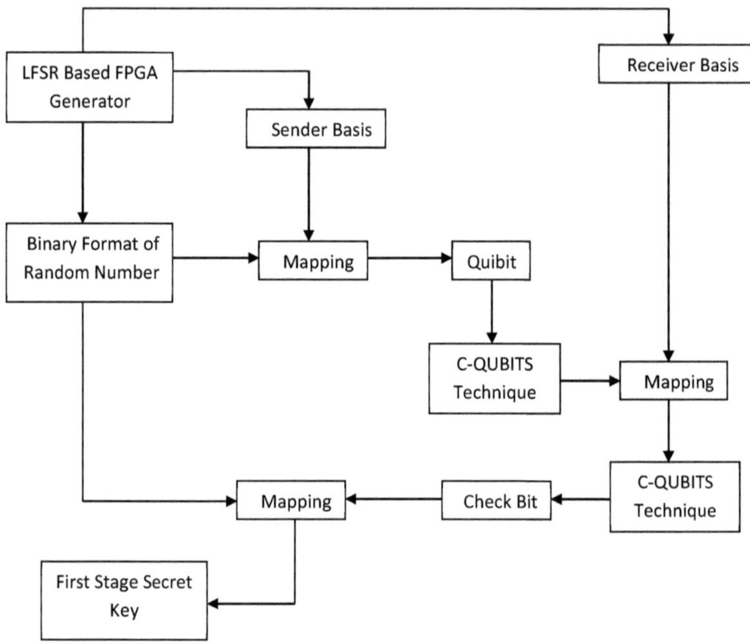

Fig. (4.9): First QKD stage implementation.

Where:

Check bits generation: receiver collects the qubits and then compares the qubits with receiver basis representation. From this, receiver generates the check bits. Then receiver sends its basis to the sender. When sender collects the receiver's quantum basis, he can generate the quantum key based on finding quantum basis match.

4.3.4 Implementing Second Stage QKD Using FPGA

Second stage QKD protocol will implement with base probability with 0.5, the implementation of this stage using FPGA is shown in Fig. (4.10) below.

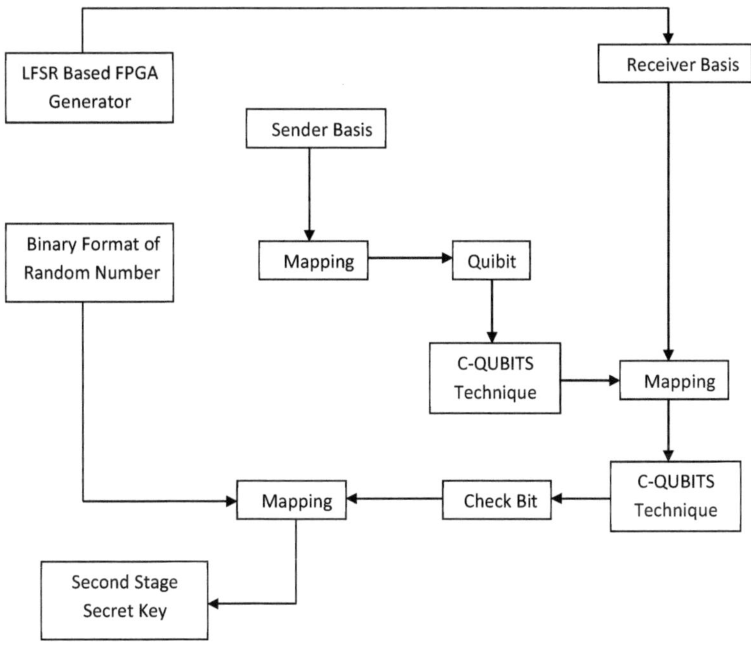

Fig. (4.10): Second QKD stage implementation.

Quantum key generation: sender got the receiver basis and then compare receiver's basis with his assumption quantum basis which is involved in generates the qubit. If both sender and receiver basis are matched from that sender generates the quantum key for secure communication as shown in Table (4.7).

Table (4.7): Quantum Key Generation.

Alice's random bits	1	1	0	1	0	1	1	0
Alice's random sending basis	+	×	×	+	×	+	+	+
Photon polarization Alice sends			\	/			/					—
Bob's random measuring basis	+	+	×	+	×	+	×	×
Photon polarization Bob measure					/			/			\	/
Bob's random bits	1	1	0	1	0	1	1	0
Shared secret key	1	—	0	1	0	1	—	—

4.3.5 C-QUBITS Technique:

C-QUBITS technique is used with BB84 QKD protocol in order to increase the security level of the whole system. In the proposed system, this technique is used with different number of bits:

a -Two-bit C-QUBITS Technique:

Two bits C_QUBITS technique includes the following steps and the actual data to code conversion is given in Table (4.8) and Fig. (4.11) shows the quantum circuit of the 2 bits C-QUBITS technique and its simulation.

i.   Alice creates a binary random number and divides them into pairs and then each pair is passed through C_NOT gate. Then it goes to Bob using randomly the two different bases + (rectilinear) and × (diagonal).

      Therefore, Alice transmits photons randomly in the four polarization states.

ii.   The bases of the pair of photons can be +×, ++, ×+, or ××.

iii.   Bob simultaneously measures the polarization of the incoming pair of photons using randomly the four possible combinations i.e. +×, ++, ×+, or ××. He does not know which of his measurements are deterministic, i.e. measured in the same pair of basis as the one used by Alice.

iv.   Later, Alice and Bob communicate to each other the list of the bases they used for each pair of photons.

v.   Bob and Alice keep only those pair of bits that were measured deterministically and will disregards those sent and measured in different bases. Statically, the pair bases coincide in 25% of all cases, and Bob's measurements agree with Alice's bits considered for begin part of the key.

vi.   Together, they can reconstitute the random bit string created previously by Alice.

Table (4.8): Actual data to code conversion.

Alice's random bits	1	1	0	1	0	1	1	0
C_NOT gate	1	0	0	1	0	1	1	1
Alice's random sending basis	+	×	×	+	×	+	+	+
Photon polarization Alice sends	|	/	/	|	/	|	|	|
Bob's random measuring basis	+	+	×	+	×	+	×	×
Photon polarization Bob measure	|	—	/	|	/	|	\	\
Bob's random bits	1	0	0	1	0	1	1	1
C_NOT gate	1	1	0	1	0	1	1	0
Shared secret key	1	—	0	1	0	1	—	—

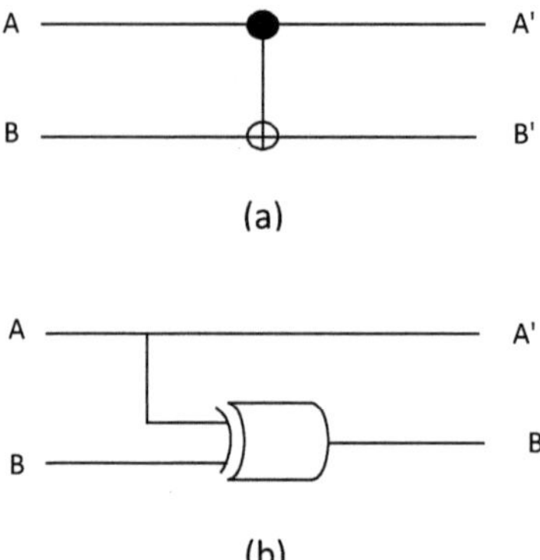

Fig. (4.11): (a) Quantum circuit of two bits C-QUBITS technique, (b) C-Qubits technique simulation.

b – Three-bit C-QUBITS Technique:

Three bits C_QUBITS technique includes the following steps and the actual data to code conversion is given in Table (4.9), Fig. (4.12) shows the quantum circuit of the 3 bits C-Qubits technique and its simulation.

vii.   Alice creates a binary random number and divides them into pairs with three bits and then each three bits is passed through C_NOT gate. Then it goes to Bob using randomly the two different bases + (rectilinear) and × (diagonal).

      Therefore, Alice transmits photons randomly in the four polarization states.

viii.   The bases of the three bits pair of photons can be +××, ++×, ×+×, ×××, +×+, +++, ×++, or ××+.

ix.   Bob simultaneously measures the polarization of the incoming pair of three photons using randomly the eight possible combinations i.e. +××, ++×, ×+×, ×××, +×+, +++, ×++, or ××+. He does not know which of his measurements are deterministic, i.e. measured in the same three basis as the used by Alice.

x.   Later, Alice and Bob communicate to each other the list of the bases they used for each three photons.

xi. Bob and Alice keep only those three bits that were measured deterministically and will disregards those sent and measured in different bases. Statically, the pair bases coincide in 12.5% of all cases, and Bob's measurements agree with Alice's bits considered for begin part of the key.

Together, they can reconstitute the random bit string created previously by Alice.

Table (4.9): Actual data to code conversion.

Alice's random bits	1	1	0	1	0	1	1	0	0
C_NOT gate	1	1	1	1	0	1	1	0	0
Alice's random sending basis	+	×	×	+	×	+	+	+	+
Photon polarization Alice sends			\	\			/					—	—
Bob's random measuring basis	+	+	×	+	×	+	×	×	×
Photon polarization Bob measure					\			/			\	/	/
Bob's random bits	1	1	1	1	0	1	1	0	0
C_NOT gate	1	1	0	1	0	1	1	0	0
Shared secret key	1	—	0	1	0	1	—	—	—

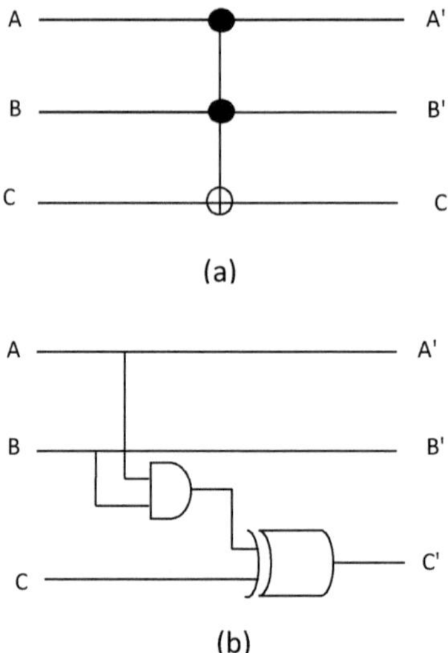

Fig. (4.12): (a) Quantum circuit of three bits C-QUBITS technique, (b) C-Qubits technique simulation.

4. 4 Transferring the Data with FPGA (Serial Transmission)

To deal with the bits inside FPGA memory, a transmission from FPGA memory to the computer is needed. Both Polarization bases and qubits must be transferred from the FPGA to the computer to simulate quantum channel transmission and then for the public channel transmission simulation. The RS232 serial port is used to transfer the data. The formatting of the data in/out in this port is shown in Fig. (4.13).

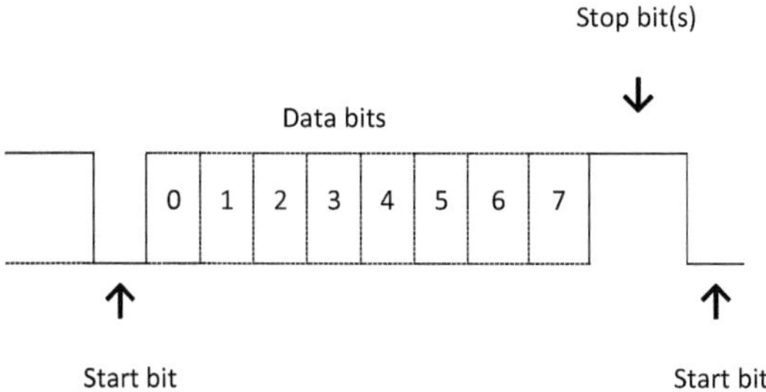

Fig. (4.13): The sequence of the transferring the data in RS232.

FPGA generates clock pulses with frequency equal to the baud rate. A counter counts a number of clocks equal to the division of the half of the frequency of the FPGA over the baud rate; then the desired clock is complemented from Low to HIGH or HIGH to LOW.

The data sent from FPGA by RS232 port will be as the state diagram shown in Fig. (4.14). When the signal (Start send) converts to HIGH and data are ready to be sent, it goes to state two; then the counter will start to count until it reaches the number of stop bits. The signal (Data OUT) will convert to zero to give a start bit and goes to state three. When it reaches to state three the signal (Data OUT) will be the valid data to send and the counter will start counting until it reaches eight; then it will go to state one and the signal (Data OUT) will be one.

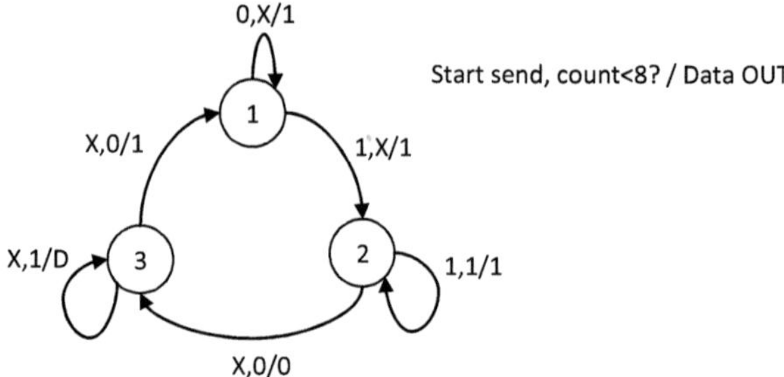

Fig. (4.14): The state diagram of output data from RS232.

Chapter Five
Results and Discussion

5.1 Background

It has already been discussed that LFSR technique is implemented using FPGA to generate the bases and the random binary bits; then these bits are transferred with serial cable. In this thesis, performance analysis of the whole system is achieved by measuring different parameters and by taking many criteria into calculations. The results of the proposed system is compared with Singh and Sharma [13] results which we measure it, and A. Kumar ,et. al. [15] results.

5.2 Quantum Measurements

Several quantum factors are measured:

1. **Raw key length:** It is the number of bits that has the same base in both sender and receiver sides; thus, these bits are received truly at the receiver side and kept as raw key. This parameter is measured and compared with raw key length of [13].

2. **Security level:** The security level performed by the length of the key caught by Eve; even though this length is short, the security level becomes high.

$$\text{Security level } \alpha \ \frac{1}{\text{number of bits caught by Eve}} \tag{5.1}$$

This parameter is compared with security level of [5] and [13] which contains the same technique (C_QUBITS) technique.

3. **Algorithm gain:** It is the gain obtained from the proposed system algorithm which is defined by Eq. (5.2)

$$\frac{\text{number of raw key bits}}{\text{total number of bits sent from Alice to Bob}} \qquad (5.2)$$

This factor is measured and compared with algorithm gain of [13].

4. **Algorithm Bit Error Rate (ABER):** It is the error rate occurred by the algorithm and it is measured by Eq. (5.3)

$$\frac{\text{number of bits with wrong bases choice}}{\text{total number of bits sent from Alice to Bob}} \qquad (5.3)$$

This factor is measured and compared with ABER of [13].

5. **Base error rate:** It performs the error rate of the base generator algorithm; the measurements are as follows:

- Rectilinear base error rate

- Diagonal base error rate

- Average base error rate which is taken as the average of rectilinear base error rate and diagonal base error rate.

This parameter is measured and compared with base error rate of [13].

6. **Probability error rate:** This is the probability of error estimated phase.

Let r_1 be the length of the first subset and r_2 is the length of the second subset just before Alice and Bob publicly compare both strings and estimate the number of errors e_1 and e_2, respectively, which leads to the error rates:

$$p_1 = \frac{e1}{r1} \qquad (5.4)$$

$$p_2 = \frac{e2}{r2} \qquad (5.5)$$

$$p = \frac{p1 + p2}{2} \qquad (5.6)$$

Where p_1 is the error rate for the photons measured with the first bases and p_2 is for those measured with the second bases while p is the average probability of error.

The average probability error rate could be measured according to the base probability δ which is:

The advantage of double error rate is as follows: If Eve starts a specific attack like the biased eavesdropping strategy, then Eve chooses the probability q_1 for measuring each photon sent from Alice to Bob in the first basis (e.g., rectilinear) and a second probability q_2 for measuring each photon in the second basis (e.g., diagonal). Hence, with probability $(1-q_1-q_2)$, Eve does not measure the photon. When Alice and Bob use the same bases, errors occur only if Eve uses a different bases. Regarding Bob's side, the photons measured by Eve are randomized and yielded an incorrect bit in half of these cases [20].

This parameter is measured and compared with [13].

5.2.1 Raw Key Length

Table (5.1) shows the raw key length of the proposed system and standard system with the same initial states of the key bits generator shift register and Fig. (5.1) shows a comparison graph of raw key lengths.

The raw key length of standard system is lower than the proposed system, since the standard system has the same base probability in both QKD stages which is 50%. The first stage of QKD in the proposed system has base probability of 60% which increases the raw key length; thus the raw key length of proposed system is higher than that of the standard system. Fig. (5.1) shows the raw key length of proposed system vs standard system. It is clear that raw key length of standard system increases slowly but in the proposed system, the length of raw key increases faster than that of the standard system; this increment becomes faster when the bits generated by Alice become more then 2000 bits.

Table (5.1): Raw key lengths.

System	Bits send by sender	Bits received truly(raw key)
Standard System	256 bits	192 bits
	512 bits	384 bits
	1024 bits	758 bits
	2048 bits	1541 bits
	4096 bits	3070 bits
Proposed System	256 bits	219 bits
	512 bits	446 bits
	1024 bits	881 bits
	2048 bits	1743 bits
	4096 bits	3469 bits

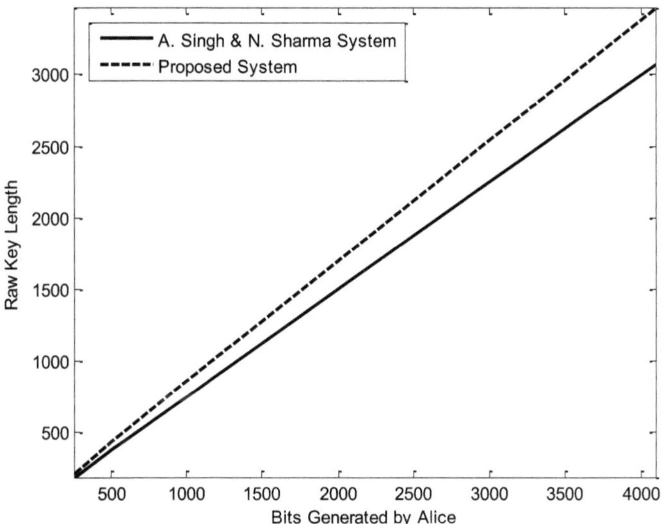

Fig. (5.1): Raw key length comparison.

5.2.2 Security Level

The security level depends on the length of key bits caught by Eve and it is dedicated by the quantum gate which is CNOT gate.

Both standard and proposed systems are tested with three cases:

Case 1: No quantum gate is added.

Case 2: with 2-bits CNOT quantum gate.

Case 3: with 3-bits CNOT quantum gate.

Fig. (5.2) shows a comparison among all the six cases and Table (5.2) shows the security level measurements. From Fig. (5.2), it could be seen that the system without quantum gate (i.e. case 1 and case 4) has low security level due to the high number of bits caught by Eve.

System with 2-bits CNOT quantum gate (i.e. case 2 and case 5) has higher security level since the probability that Eve caught the bits with right state is decreased to $\frac{1}{4}$.

Finally, the lowest number of bits caught by Eve is at the system with 3-bits CNOT quantum gate (i.e. case 3 and case 6) since Eve's probability to catch bits is decreased to $\frac{1}{8}$.

The comparison in Fig. (5.2) shows that the proposed system with case 6 (i.e. with 3-bits CNOT quantum gate) is the best case.

The lowest line in Fig. (5.2) is the line performing the bits caught by Eve of the proposed system/case 6.

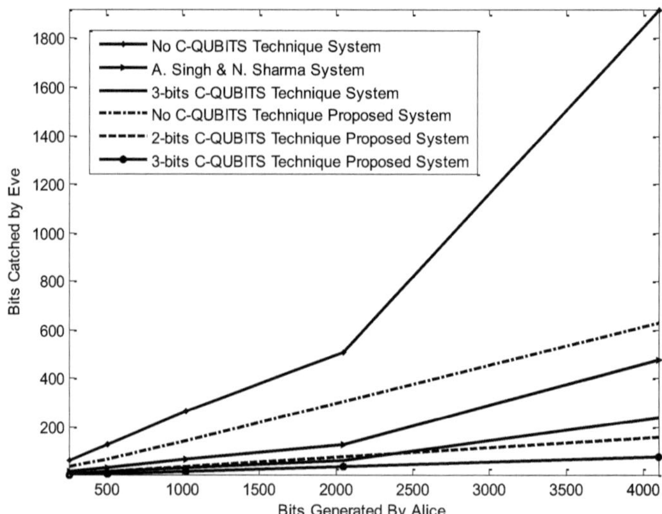

Fig. (5.2): Bits caught by Eve comparison.

Table (5.2): Theoritical calculations of security level.

Case #	Bits send by sender	Bits cached by Eve
Case 1	256 bits	64 bits
	512 bits	128 bits
	1024 bits	266 bits
	2048 bits	507 bits
	4096 bits	1921 bits
Case 2	256 bits	16 bits
	512 bits	32 bits
	1024 bits	67 bits
	2048 bits	127 bits
	4096 bits	480 bits
Case 3	256 bits	8 bits
	512 bits	16 bits
	1024 bits	33 bits
	2048 bits	63 bits
	4096 bits	240 bits
Case 4	256 bits	37 bits
	512 bits	66 bits
	1024 bits	143 bits
	2048 bits	305 bits
	4096 bits	627 bits
Case 5	256 bits	9 bits
	512 bits	17 bits
	1024 bits	36 bits
	2048 bits	76 bits
	4096 bits	157 bits
Case 6	256 bits	5 bits
	512 bits	8 bits
	1024 bits	18 bits
	2048 bits	38 bits
	4096 bits	78 bits

5.2.3 Algorithm Gain

Algorithm gain of the proposed system is higher than algorithm gain of the standard system as shown in Table (5.3).

The results of the comparison in Fig. (5.3) demonstrate that the proposed system is better than the standard system in algorithm gain. The algorithm gain of the proposed system is approximately constant after point 1500 and is higher than the algorithm gain of the standard system which is decreased so fast and starts dropping after point 2000.

Table (5.3): Algorithm gain.

System	Bits send by sender	Algorithm Gain
Standard System	256 bits	75%
	512 bits	75%
	1024 bits	74.0234%
	2048 bits	75.2441%
	4096 bits	74.9512%
Proposed System	256 bits	85.5469%
	512 bits	87.1094%
	1024 bits	86.0352%
	2048 bits	85.1074%
	4096 bits	84.6924%

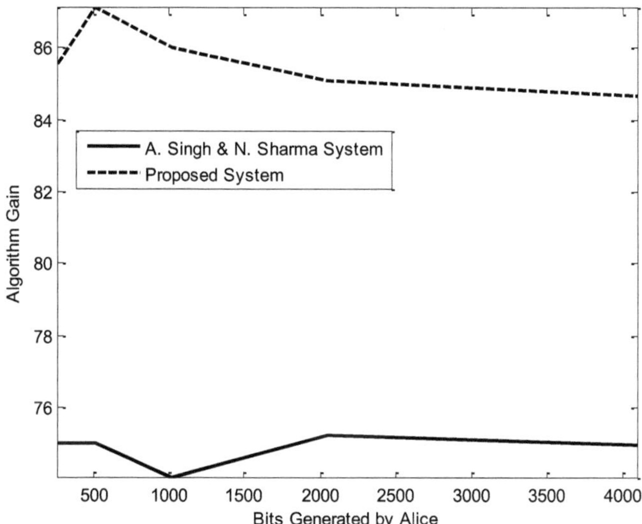

Fig. (5.3): Algorithm gain comparison.

5.2.4 Algorithm Bit Error Rate

Algorithm bit error rate is generated due to the unmatching bases of the receiver side with the bases chosen at the sender side.

Table (5.4) shows the algorithm bit error rate to the system of both standard system and proposed system. The comparison is shown in Fig. (5.4).

Table (5.4): Algorithm bit error rate (ABER).

System	Bits send by sender	Algorithm bit error rate (ABER)
Standard System	256 bits	25%
	512 bits	25%
	1024 bits	25.9766%
	2048 bits	25.7559%
	4096 bits	46.8994%
Proposed System	256 bits	14.4531%
	512 bits	12.8906%
	1024 bits	13.9648%
	2048 bits	14.8926%
	4096 bits	15.3076%

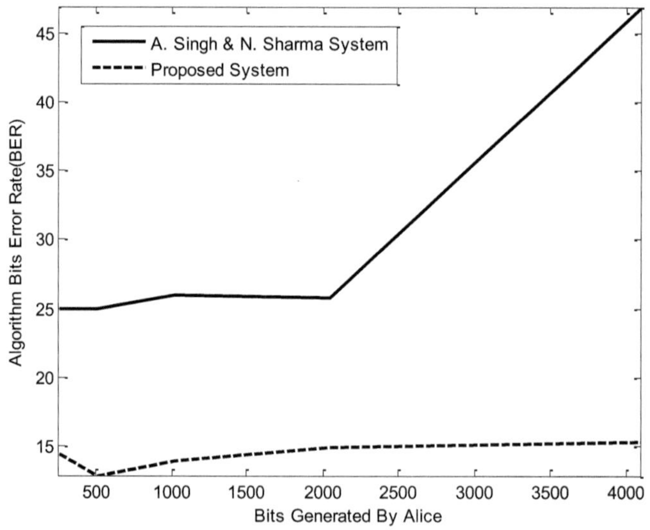

Fig. (5.4): Algorithm bit error rate comparison.

From Fig. (5.4), it clearly could be seen that the proposed system is better in algorithm bit error rate factor than the standard system since the algorithm bit error rate of the proposed system in its three cases is almost linear after its slight increase after 500 bits; while the standard system algorithm bit error rate is a linear line at points 500, 1000, 1500, and 2000 bits and then it starts increasing sharply after 2000 bits.

5.2.5 Base Error Rate

The following table contains the measurements of the base error rate which were taken to rectilinear base, diagonal base, and the average base error rate of the all base.

Table (5.5): Base error rate measurements.

Bits send by sender	Standard System			Proposed System		
	Rectilinear base error rate	Diagonal base error rate	Average base error rate	Rectilinear base error rate	Diagonal base error rate	Average base error rate
256 bits	50.39%	50%	50.195%	50.51%	33.62%	42.065%
512 bits	50.2%	49.5%	49.85%	50.75%	36.48%	43.625%
1024 bits	50.1%	48.7%	49.4%	48.86%	35.59%	42.225%
2048 bits	50.05%	47.3%	48.675%	49.58%	35.71%	42.645%
4096 bits	50.02%	46.8%	48.41%	49.21%	35.59%	42.4%

From Fig. (5.5) which is showing a comparison graph of the average base error rate between the standard system and the proposed system, it is clear that the proposed system is better than the standard system. In proposed system, base probability of second stage is 0.6 as mentioned before; this increases the probability of matching bases choice at sender and receiver sides thus, from Fig. (5.5) the proposed system base error rate line is lower than the standard system base error rate line and it is almost linear after 1500 bits.

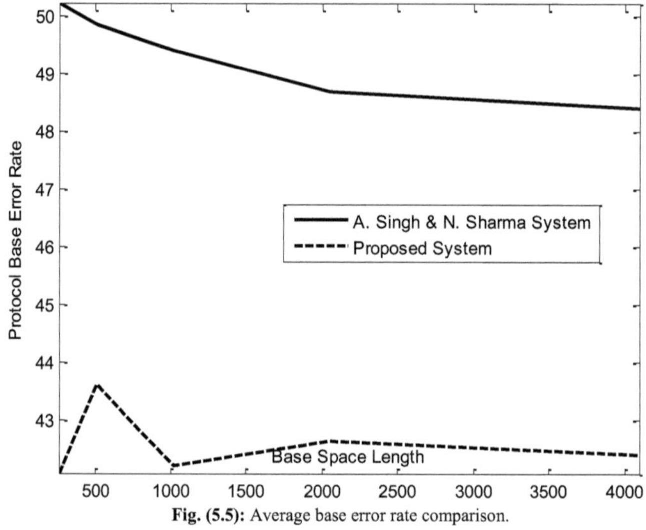

Fig. (5.5): Average base error rate comparison.

5.2.6 Probability Error Rate

Table (5.6) shows the measurements of probability error rate for both standard system and proposed system. A comparison between the proposed system probability error rate and the standard system probability error rate is shown in Fig. (5.6). From this figure it is clear that the proposed system is better than the standard system since it has a probability of error slicely lower than that of the standard system.

Table (5.6): Probability error rate measurements.

System	Bits send by sender	Probability Error Rate
Standard System	256 bits	4.1829%
	512 bits	1.2782%
	1024 bits	0.3529%
	2048 bits	0.1803%
	4096 bits	0.1274%
Proposed System	256 bits	3.5054%
	512 bits	1.1183%
	1024 bits	0.3016%
	2048 bits	0.1579%
	4096 bits	0.1116%

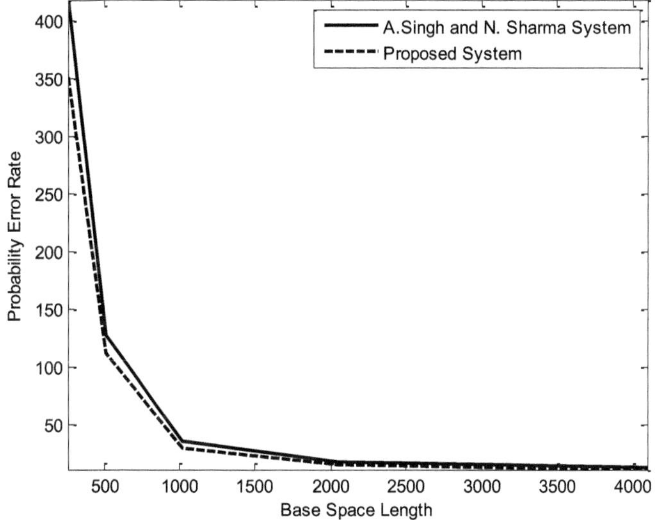

Fig. (5.6): Probability error rate comparison.

5.3 FPGA Performance Analysis

In this thesis, FPGA kit is used to generate the binary bits which are polarized later and transmitted from the sender side to the receiver side and also used to generate binary presentation to the polarizer bases (i.e. rectilinear base with logic '0' and diagonal base with logic '1'); LFSR algorithm is used as FPGA generator. So, FPGA performance measurements are done according to the following factors:

1. **Time to complete the total state:** Any algorithm implementation on FPGA will be done as states and each state require a specific amount of time to complete its building.

2. **Number of states generating:** This factor calculates the number of the states in the algorithm.

3. **Clock:** It is the clock period of FPGA device; this clock is taken from external crystal with a predefined frequency according to the FPGA type and size.

4. **Shift register:** It performs a number of initial bits entered to the FPGA device as initial states to the LFSR register and this will perform the length of the LFSR register implemented over FPGA.

5. **XOR gate:** This is the number of XOR gate required to implement the algorithm.

6. **Time to transfer one bit:** In the proposed system, the bits and bases need to be taken out of the FPGA (i.e. out from FPGA memory to the computer) and this is done by using serial cable (USB to RS232) and this transfer requires a specific amount of time to be completed; time to transfer one bit from the FPGA to a computer over USB to RS232 cable is measured in this factor.

7. **Transfer time (delay):** It is the total time required to transfer all the data from the FPGA to the computer and it is considered as a delay to the operation.

8. **Total pin:** It is the number of the total FPGA pins used in the proposed systems measured in this factor.

Table (5.7) shows the performance of FPGA, and a comparison is done between this thesis FPGA measurement and the measurement of FPGA performance of [15]; Figures (5.7), (5.8), and (5.9) show some features of comparison in which the enhancement is done.

Table (5.7): FPGA performance analysis.

Performance analysis	8 bits LFSR	9 bits LFSR	10 bits LFSR	11 bits LFSR	12 bits LFSR	16 bits LFSR	32 bits LFSR
Time to complete the total states	< 20 ns	< 20 ns	< 20 ns	< 20 ns	< 20 ns	< 20 ns	< 20 ns
Clock	20 ns	20 ns	20 ns	20 ns	20 ns	20 ns	20 ns
Shift register	08	09	10	11	12	16	32
XOR gate	01	01	01	01	01	01	01
(transfer time) Delay	2.2 ms	4.4 ms	8.9 ms	1.78 ms	3.55 ms	0.5689 s	3.7283 s
Total pin	03	03	03	03	03	03	03

Time to complete total states is less than 20 ns, because the LFSR implementation over FPGA is done without clock, so the LFSR algorithm will be done as a single state and to complete this state, the time will be less then one clock period which is 20 ns.

Since the LFSR algorithm is done as a single state so the total no. of random states generating will be 01.

Crystal frequency of the Altera Cyclone IV E 4CE115 FPGA is 50 MHz so its clock period will be $\frac{1}{50\ MHz}$ which is 20 ns.

Shift register is the length of LFSR register and length of the initial bits entered to the FPGA internally as initial state.

XOR gate required to generate the bit sequence using LFSR is one XOR gate to all LFSR register lengths since this XOR is xor the feedback tap of the register.

The baudrate of the serial transmission of cable USB to RS232 is 115200 bps; this makes the time to transfer one bit be $\dfrac{1}{115200 \text{ bps}}$ and it is equal to 8.6806 µspb.

The delay for this thesis proposed work was the time spent to transfer operation and this time is the time taken by all bits to transfer from FPGA to computer device and it is equal to the time required to transfer one bits multiplied by the number of actually bits required to be transferred.

Number of pins refers to how many pins used of FPGA to implement the LFSR generator; in this proposed work, it is 3 pins which are:

- ❖ PIN_Y2: It is the clock pin and it is connected to the CLK generator.
- ❖ PIN_AB28: It is the input switch used to start the transfer operation when it is logic '1'.
- ❖ pin_g9: It is the serial pin and it is connected to serial transfer circuit to start the transfer operation.

Fig. (5.7) shows the time required to complete the total states; in this factor, the proposed method is better since its time is less than the time of A. Kumar method. Time to complete the total state of A. Kumar proposed method will be increased linearly after the shift register being more than 15 bits. Thus, this thesis proposed method is better since the time to complete the total states remain constant no matter how long the shift register is (i.e. the proposed method is much faster than A. Kumar method).

Fig. (5.8) shows the number of random states generated during the generation process. This thesis proposed system is better than A. Kumar method because this proposed system produced one state since this proposed system states are executed as a single state at the same time due to that the generation process is done without a clock; the clock was used for the transfer process.

The number of generated random states of the proposed method shown in Fig. (5.8) will stay constant due to that the LFSR algorithm is used in proposed method and executed without a clock, thus the output will be generated directly with one state and that makes the proposed method better than A. Kumar method whose random state increased linearly with the increase of the shift register length.

The proposed method used just three pins of the FPGA kit whatever the shift register length is; this makes the proposed method better than A. Kumar method which uses more pins when shift register length increases; it started with 10 pins at 8 bit LFSR algorithm and increased as the LFSR register length increased as shown in Fig. (5.9).

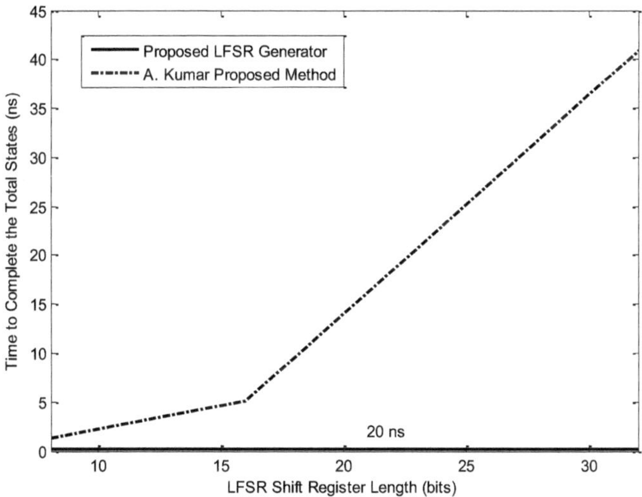

Fig. (5.7): Time to complete the total states comparison.

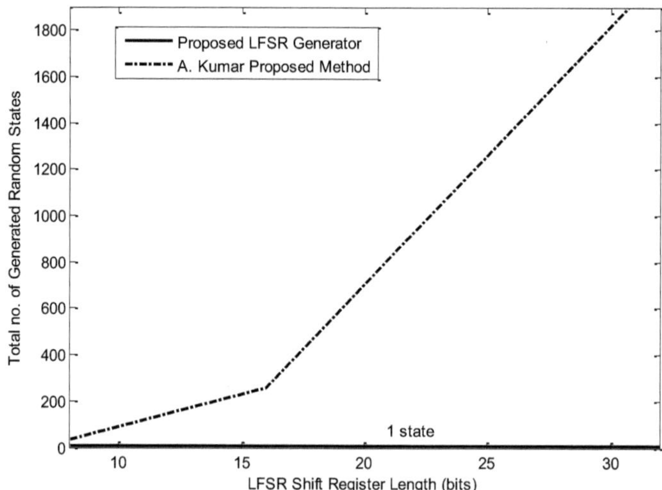

Fig. (5.8): Random states number comparison.

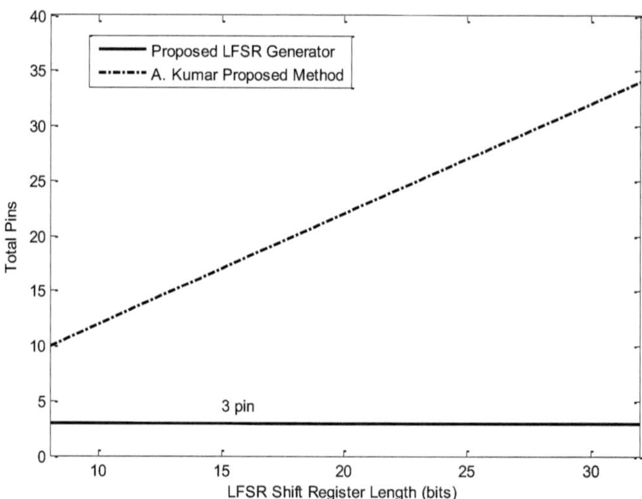

Fig. (5.9): Total pins number comparison.

Chapter Six

Conclusions and Future Works

6.1 Conclusions

The proposed model implements Quantum cryptographic bit and base generation algorithm in FPGA. The enhancements made over this thesis system were in two quantum protocol stages: first, sifting key stage and second, error elimination and correction stage.

Several points were concluded throughout the building of the proposed circuits and analyzing the overal QKD system and FPGA circuits can be summarized by the following:

- By increasing the QKD steps, the raw key length increases.
- In the increment in the base probability factor δ, matching basis at both sender and receiver sides raise will increase the raw key length and the protocol gain.
- By adding a quantum gate to the transmitting system, maximum probability of catching bits by eavesdropping will decrease to $\frac{1}{4}$ if two bits quantum gate is used and to $\frac{1}{8}$ if three bits quantum gate is used; this increases the security level.
- Entering most of the FPGA simulation parameters internally will speed up the execution process and decrease the hardware level of FPGA (i.e. the number of used FPGA pins will decrease).

❖ Decreasing the number of random executable states by executing LFSR algorithm without clock pulses will speedup the time required to get the generating output.

6.2 Future Works

Suggestions for future research can be summarized by some points:

❖ Increasing QKD security level by making enhancements in other QKD protocol stage like privacy amplification stage.

❖ After that, QKD system becomes completely synthesized; it can be connected to a cryptographic system like RSA cryptographic system to encrypt and decrypt different kinds of messages to obtain a complete ciphering/deciphering security system.

❖ Although it is known to use BB84 protocol as a QKD protocol, there are some applications that use SARG04 protocol for implementing QKD purposes; therefor, a SARG04 protocol implementation can be suggested a future research.

❖ Implementing QKD network over the FPGA as a self dependent Integrated Circuit (IC) without the need to use a PC (i.e. stand alone QKD –FPGA_ based system).

References

[1] S. Tayeh, " Error Elimination and Privacy Amplification in Quantum Cryptosystems", Ph.D. Dissertation, University of Al_Nahraien, Iraq, 1999.

[2] C. H. Zhu, D. X. Quan, F. Zhang, and C. X. Pei, "Improving Key Rate of Optical Fiber Quantum Key Distribution System Based on Channel Tomography", International Journal of Theoretical Physics, Springer, Vol. 52, P.P. 596-603, 2013.

[3] Portland Quantum Logic Group, "Evolving Quantum Circuits and an FPGA Based Quantum Computing Emulator", Advanced elecommunications Research Institute International, Japan, 2002.

[4] P. Navez and G. V. Assche, " A method for secure transmission Quantum Cryptography", Technical IT_Scan, 2002.

[5] D. Singh and A. Singh, "An Effective Technique For Data Security in Modern Cryptosystem", BVICAM'S International Journal of Information Technology (BIJIT), Vol. 2, No. 1, ISSN 0973 – 5658, 2010.

[6] H. Abbas, "Quantum Mixed Transform", M.Sc. Thesis, University of Baghdad, 2011.

[7] H. Ansari, A. Parameswaran, L. Antani, B. Aditya, A. Taly, and L. Kumar, "Quantum Cryptography and Quantum Computation", Department of Computer Science and Engineering, Indian Institute of Technology, 2006.

[8] E. Al-Daoud, " Comparing Two Quantum Protocols: BB84 and SARG04", European Journal of Scientific Research, Vol. 17, No. 1, ISSN 1450-216X, 2007.

[9] C. J. Ware, "Modeling and Analysis of Quantum Cryptographic Protocols", M.Sc. Thesis, University of Victoria, Department of Computer Science, 2008.

[10] M. S. Sharbaf, " Quantum Cryptography: A New Generation of Information Technology Security System", Graduate School of Computer and Information Sciences, Nova Southeastern University, Fort Lauderdale, Fl., 33314, Sixth International Conference on Information Technology: New Generations, 2009.

[11] N. A. Muhammad and Z. A. Zukarnain, "Implemetation of BB84 Quantum Key Distribution Protocol's with Attacks", European Journal of Scientific Research, Vol.32, No.4 , ISSN 1450-216X, pp.460-466, 2009.

[12] N. Kaur, A. Singh, and S. Singh, " Enhancement of Network Security Techniques using Quantum Cryptography", International Journal on Computer Science and Engineering (IJCSE), Vol. 3, No. 5, ISSN: 0975-3397, 2011.

[13] A. Singh and N. Sharma, "Development of Mechanism for Enhancing Data Security in Quantum Cryptography", Advanced Computing: An International Journal (ACIJ), Vol.2, No.3, 2011.

[14] V. A. Devi and T. S. Raj, "Quantum Cryptography Based Email Communication Through Internet", International Journal of Engineering Science and Technology (IJEST), Vol.3, No.1, ISSN: 0975-5462, 2011.

[15] A. Kumar, P. Rajput, and B. Shukla, " Design of Multi Bit LFSR PNRG and Performance Comparison on FPGA using VHDL", International Journal of Advances in Engineering & Technology, ISSN: 2231-1963, 2012.

[16] J. Preskill, " Quantum computing: pro and con", The royal society, 2012.

[17] T. Huffmire, C. Irvine, T. D. Nguyen, T. Levin, R. Kastner, and T. Sherwood, "Handbook of FPGA Design Security" springer, ISBN 978-90-481-9156-7, e-ISBN 978-90-481-9157-4, 2010.

[18] L. Vincent, " Practical Security of Quantum Cryptography", Ph.D. Dessiertation, Norwegian University of Science and Technology Faculty of Information Technology, 2011.

[19] S. Imre and F. Bal´azs, " Quantum Computing and Communications An Engineering Approach", John Wiley and sons, ISBN 0-470-86902-X.

[20] C. Kollmitzer and M. Pivk, "Applied Quantum Cryptography", (Springer, Berlin Heidelberg), ISSN 0075-8450, 2010.

[21] M. Thomas, " Quantum Communication and Teleportation Experiments using Entangled Photon Pairs", Ph.D. Dissertation, University of Wien, 2002.

[22] Y. S. Atiya, "Quantum Applications in OFDM System", M.Sc. Thesis, University of Bghdad, 2012.

[23] M. A. Mehdi, "Performance Improvement of Elliptic Curve Cryptography Using FPGA", M.Sc. Thesis, University of Bghdad, 2012.

[24] R. Wain, I. Bush, M. Guest, M. Deegan, I. Kozin, and C. Kitchen, "An Overview of FPGAs and FPGA Programming; Initial Experiences at Daresbury", CCLRC Daresbury Laboratory, Version 2.0, Daresbury, Warrington, Cheshire, WA4 4AD, UK, 2006.

[25] N. Chhedaiya and V. Moyal, " Implementation of Back Propagation Algorithm in Verilog", Int.J.Computer Technology & Applications,Vol. 3 (1), PP. 340-343, ISSN:2229-6093, 2012.

[26] M. M. Mano and M. D. Ciletti," Digital Design", fourth edition,ISBN: 0-13-234043-7, Pearson prentice hall, upper saddle river, NJ 07458.

[27] G. R. Smith, "FPGAs 101: Everything you need to know to get started", ISBN 978-1-85617-706-1 (Elsevier, USA), 2010.

[28] I. Grout, "Digital Systems Design with FPGAs and CPLDs", Elsevier Ltd., 2008.

[29] W. Wolf, "FPGA-Based System Design", Upper Saddle River, NJ 07458, ISBN: 0-13-142461-0, 2004.

[30] M. Bhatt, A. Aneja, and S. Tripathi, "Classical Cryptography v/s Quantum Cryptography A Comparative Study", International Journal of Electronics and Computer Science Engineering, ISSN-2277-1956, 2012.

[31] D. Hrg, L. Budin, and M. Golub, "Quantum Cryptography and Security of Information Systems", University of Zagreb, Faculty of Electrical Engineering and Computing, Zagreb, 2002.

I want morebooks!

Buy your books fast and straightforward online - at one of world's fastest growing online book stores! Environmentally sound due to Print-on-Demand technologies.

Buy your books online at
www.morebooks.shop

Kaufen Sie Ihre Bücher schnell und unkompliziert online – auf einer der am schnellsten wachsenden Buchhandelsplattformen weltweit! Dank Print-On-Demand umwelt- und ressourcenschonend produziert.

Bücher schneller online kaufen
www.morebooks.shop

Printed by Books on Demand GmbH, Norderstedt / Germany